PENGUIN BOOKS

THE CRASH DETECTIVES

CHRISTINE NEGRONI is a journalist specializing in air travel and aviation for *The New York Times*, ABC News, *Air & Space*, *The Huffington Post*, and many other news organizations. She began her career in broadcasting as an anchorwoman in local television and rose to become a network correspondent at CBS News and CNN. While working in CNN's New York bureau, she covered the crash of TWA Flight 800 and went on to write the book *Deadly Departure*, which was a *New York Times* Notable Book of the Year. Neither a pilot nor an engineer, she was nonetheless asked by the Federal Aviation Administration to represent the traveling public and present a fresh-eyed perspective on a five-year advisory committee formed to address problems exposed by the crash of TWA Flight 800 and the 1998 in-flight fire on Swissair Flight 111, which killed all on board.

Following the attacks against America on September 11, 2001, she joined the aviation law firm Kreindler & Kreindler, directing its investigation into sponsorship of terror and other aviation disasters on behalf of victims' families. During this time she qualified for membership in the International Society of Air Safety Investigators.

She is considered a thought leader in the aviation industry and contributes insight, analysis, and advocacy on the subjects of safety and civility in air travel.

THE CRASH DETECTIVES

INVESTIGATING THE WORLD'S MOST MYSTERIOUS AIR DISASTERS

CHRISTINE NEGRONI

PENGUIN BOOKS

PENGUIN BOOKS

An imprint of Penguin Random House LLC
375 Hudson Street
New York, New York 10014
penguin.com

LIBRARY OF CONGRESS CATALOGING-IN-PUBLICATION DATA

Names: Negroni, Christine, author.
Title: The crash detectives : investigating the world's most mysterious air disasters / Christine Negroni.
Description: First edition. | New York, New York : Penguin Books, [2016]
Identifiers: LCCN 2016016761 | ISBN 9780143127321 (pbk.)
Subjects: LCSH: Aircraft accidents—Investigation. | Aircraft accidents—Human factors. | Aeronautics—Safety measures.
Classification: LCC TL553.5 .N339 2016 | DDC 363.12/465—dc23

Printed in the United States of America
1 3 5 7 9 10 8 6 4 2

Set in Sabon LT Pro • *Designed by Elke Sigal*

CONTENTS

O Trinity of love and power
All travelers guard in danger's hour;
From rock and tempest, fire and foe,
Protect them whereso e'er they go;
Thus evermore shall rise to Thee
Glad praise from air and land and sea.

—WILLIAM WHITING, 1860
PRESBYTERIAN HYMNAL

INTRODUCTION

This I can say about Malaysia Airlines Flight 370: there is little to suggest the pilots were involved in hijacking or crashing the plane they were flying from Kuala Lumpur to Beijing on March 8, 2014. One need only look at the shocking, intentional crash of a GermanWings* flight one year later to see how quickly, and how many, clues emerge when a pilot plots to bring down an airliner. My theory about what happened to MH-370 began to form within a week of arriving in Malaysia to help ABC News cover the story.

When I first heard about the missing flight, I was at sea in Vietnam's Gulf of Tonkin. The fact that the news reached me in such a remote place was a new high in communications technology. That years later we do not know what happened to the airliner and its passengers demonstrates the shallows.

I hurried to Kuala Lumpur and spent five weeks there. Each night, I went to bed pretty sure I'd wake up to hear that

---

* A subsidiary of Lufthansa, GermanWings became Eurowings in 2016.

the airliner had been found. When it was not, I was swept along with everyone else in believing that this was "unprecedented," as Malaysia's transportation and defense minister was fond of saying.

In fact, over the past century of commercial aviation, more than a dozen airliners have disappeared without a trace. And even when a missing plane is found, it is sometimes impossible to determine what went wrong.

When I returned to the United States and started the research for this book, I came across the trailer for a documentary produced by Guy Noffsinger, a media specialist at NASA. "What happened to the most high-tech commercial airliner in the world and those people aboard it?" the narrator asked ominously. Was it structural failure, pilot error, or something more sinister?

In a similar vein, author Edgar Haine in *Disaster in the Air*, writes, "Of particular concern to everyone was the sudden termination of normal radio contact" and "the absence of subsequent communications."

Yet Noffsinger and Haine weren't referring to MH-370; they were talking about the Pan American Airways flying boat the *Hawaii Clipper.** It disappeared seventy-six years before MH-370 and was one of the first mysteries in commercial aviation. It remains a subject of fascination to this day.

After two decades writing about air safety and working as an accident investigator, I have learned that most accidents are variations on a limited number of themes, and in this book I explore some of them: communication failures, overreliance on or misunderstanding of technology, errors in the design of

* Pan American Airways became Pan American World Airways in 1950.

airplanes and engines, and lapses in the performance of flight crews, operators, and mechanics. The tie that binds the accidents (and incidents*) in this book is that they confounded the crash detectives looking for answers.

Why conduct investigations anyway? It is not to provide "closure" for families of victims, though that's a compassionate side benefit. It is not to assign blame so people can be prosecuted and lawyers can sue. Investigations help illuminate how machines and humans fail, which in turn shows us how to prevent similar events. Because the aviation community has been so conscientious about this over the years, hurtling through the air at five hundred miles an hour and seven miles high is far less likely to kill you than almost any other type of transportation.

From pilot training and airplane and engine design to dropping crash test dummies on their rubberized and sensor-equipped bottoms to the floor of a test lab, every decision in commercial aviation is based on lessons learned from disaster. That's why it is so important to discover what happened to Malaysia 370, even if the plane is never found.

An unsuccessful search is still not the end of the story; thinking through scenarios of what might have happened can identify hazards that need to be fixed. So while it is possible that one or both of the pilots—in an uncharacteristic act of hostility and without any of their friends or family noticing anything amiss—purposefully took the plane on a flight into oblivion, other theories better fit the available facts.

My theory is that an electrical malfunction knocked out systems on the Boeing 777 and that the plane lost pressurization,

---

* Incidents do not involve serious injury, loss of life, or substantial damage to the airplane.

incapacitating the pilots. Whatever happened, it could not have caused damage serious enough to affect the airworthiness of the plane, since it flew on until running out of fuel many hours later. Likely, the men in the cockpit were overcome by the altitude sickness known as hypoxia, which robbed them of the ability to think clearly and land the plane safely. Many of the links in the bizarre chain of events that night can be explained by hypoxia because past cases have shown how rapidly those who fall victim to it turn imbecilic.

As soon as a plane crashes, people begin to speculate about what happened. Horace Brock, who became a Pan Am pilot shortly after the *Hawaii Clipper* disappeared, noted in his book, *Flying the Oceans*, "The public will not tolerate a mystery. They always sense a conspiracy. They will never believe in coincidence or even a predictable tragedy."

Alternative theories abound in many notable accidents, including the disappearance of Amelia Earhart in 1937, the death of United Nations secretary-general Dag Hammarskjöld in Northern Rhodesia* in 1961, and the 1996 in-flight explosion of TWA Flight 800 off the coast of New York.

Questioning the official version of events can be a good thing. The loss of an Air New Zealand DC-10 on a sightseeing flight over Mount Erebus in Antarctica in 1979 was first attributed to pilot error. Only after people outside the investigation presented their own evidence did a special court of inquiry discover what it called "a litany of lies" by an airline and a government trying to hide their culpability. More on that crash later.

The tradition of doubt in aviation goes back to Orville and Wilbur Wright's successful first flights, which prompted an edi-

---

* Now Zambia.

torial writer to say of the brothers three years later, "They are in fact either fliers or liars. It is difficult to fly. It is easy to say, 'We have flown.'"

If skepticism was a gnawing mouse in flying's early days, it is a roaring lion now that anyone with an Internet connection can access information and review the evidence. Invited or not, independent analysts and armchair investigators are contributing to the discussion on TV news, blogs, and pop-up crowdsourcing sites. For the first time in history, technology is connecting hyper-specialists with geeks, skeptics, and advocates. Information can be scrutinized and analyzed in ways not previously possible, and this Internet-enhanced coalescing of the world's brain power will certainly continue to grow.

This book is a part of that evolution as I hypothesize about MH-370 and other disasters that have mystified the world.

PART ONE

Mystery

I have approximate answers and possible beliefs and different degrees of uncertainty about different things.

—NOBEL PRIZE–WINNING THEORETICAL PHYSICIST RICHARD FEYNMAN

The Clipper

On the last leg of a journey halfway around the world, Pan American Airways captain Leo Terletsky began to worry. And when Captain Terletsky worried, everybody else on the flight deck worried, too. "His anxiety caused him to shout at copilots, issue orders and immediately countermand them. He infected his crews with his own anxiety," wrote Horace Brock, who flew a few times with Terletsky and didn't much like it.

At noon on July 29, 1938, there was plenty to be anxious about. Having spent fifty-six hours over five days flying Pan Am passengers from San Francisco to the Far East, Terletsky and his nine-man crew were deep into bad weather as the Martin 130 flying boat approached the Philippine archipelago.

The plane was "sandwiched between two layers of clouds," explained Pete Frey, a captain with a large American carrier and a safety investigator with his union, who reviewed for me the weather reports submitted by the crew on that miserable summer morning. The stratocumulus clouds that Terletsky encountered are often at the beginning or end of worse weather, including rain and turbulence. Terletsky was dealing with both as he threaded the four-engine plane through the bands of clouds above and below where he was flying at 9,100 feet, 586 miles east of Manila. As Frey explained it to me, the rocky ride was not the crew's biggest problem.

"They are inside the clouds half the time, flying on instruments. This would make navigating by observing landmarks below impossible. It would also make getting a fix from the sun or other celestial object impossible as well.

"They are navigating with dead reckoning," Frey said. Dead reckoning is the most basic form of navigation: essentially a mathematical calculation involving weather, wind, time, speed, and direction. "You make an assumption of wind correction and simply hold a heading and course for a fixed period of time. At the end, you hope to be where you planned," Frey explained. Yet considering their inability to see the earth below, the crew would have had little on which to base their position; or as Frey imagined flying under the circumstances, "You are lost."

Around noon local time, radio operator William McCarty, thirty-three, sat at his desk behind the copilot, tapping the keys of his Morse code machine. He was sending a message to the Pan Am ground station on the Philippine island of Panay. Even if the crew was uncertain of its position, Pan Am's ground personnel would try to use radio wave direction finding to pinpoint the flying boat's location. They could also provide the crew with information about the weather ahead.

McCarty reported the weather and the winds, the temperature, and the crew's approximation of where they were, along with their speed. Morse code could get through even when the plane's radio signal was not strong enough to transmit a voice. By the time McCarty was done, about ten minutes had passed, and Edouard Fernandez, the operator at Panay, wanted to pass the weather news on to the crew. McCarty asked him to wait. "Stand by for one minute before sending as I am having trouble with rain static." When Fernandez tried later to contact the Clipper, there was no reply.

Nothing was ever again heard from the *Hawaii Clipper.* No piece of the plane, no human remains, no luggage or cargo, and no airplane fluid or fuel would show up. As with Malaysia 370 seventy-six years later, only the evidence still on the ground would be available for investigators to consider. They could scour the maintenance records and operational history of the plane and review the performance and training of the crew along with the information sent by McCarty during the flight, but it might not be enough to determine conclusively what happened. It could be illuminating; it might be baffling. It turned out to be both.

Thin Air

No one knows for sure what happened aboard Malaysia 370. The scenario I am about to describe is based on a framework of events put forward by Malaysian and Australian investigators and other sources who participated in gathering or analyzing the known data. To this I have applied Occam's razor, the principle that suggests that if there are many possible explanations for something, the simplest is the most likely.

Shortly after midnight on March 8, 2014, and seemingly without warning, what had been an entirely normal flight devolved into an illogical series of events. That kind of wacky has been seen before when pilots are afflicted with altitude sickness, known as hypoxia.

An inability to get enough oxygen into the lungs to sustain cogent thought happens when planes lose pressurization, and that can happen for a variety of reasons. It can be triggered by an electrical problem or some mechanical difficulty. Pilots sometimes fail to turn the pressurization on at the beginning of the flight, but even when the pressurization is working as it should, there's no way to keep a plane pressurized if there is a hole in the fuselage or if leaks at the seals of doorways, windows, or drains from the galley and bathrooms allow the denser air to escape.

If the pilots on Malaysia Flight 370 experienced oxygen deprivation because something happened to cause the plane to lose pressurization, they would have behaved irrationally, perhaps turning a moderate problem into a catastrophic one. The passengers and crew would have become feebleminded and helpless.

At the time of the MH-370 disaster, people were boarding airplanes around the world at a rate of eight million a day. Few air travelers then (or now) gave a thought to the fact that outside those aluminum walls the air is too thin to sustain coherent thought for more than a few seconds. Life itself is extinguished in minutes. While the percentage of oxygen in the air (21 percent) is the same as on the ground, the volume of air expands at higher altitudes. We rely on air density for the pressure that drives oxygen into our bodies. Miles above the earth and absent this pressure, oxygen will rush out like air racing out of a balloon.

What keeps us air travelers alive and, for the most part, in our right minds is a relatively simple process that pumps air into the plane as it ascends, like air filling a bicycle tire. The air comes off the engines and is distributed via ducts throughout the plane. In most airliners, the cabin pressure is set to mimic the pressure density of about eight thousand feet. So to your body, flying is like being in Aspen, Colorado, or Addis Ababa, Ethiopia.

When it is time to land, the valves that closed on takeoff to maintain that air density in the cabin begin to open, allowing it to escape gradually until the inside of the plane is equalized with the outside, or generally, to 190 feet above the altitude of the airport. You'll know this process is happening when your ears start to pop in the last twenty to thirty minutes of your flight. If this extra pressure weren't allowed to vent, the door of the plane might explode outward. It happened as recently as 2000, when an American Airlines Airbus A300 made an emergency landing at Miami International. Insulation blankets blocked the outflow valves, so the differential pressure inside the cabin was still high even after landing. It is not clear if the flight attendants realized it, because they had other problems. A smoke alarm had triggered, and they were worried about a fire. So they were trying to evacuate the plane, but the doors would not budge. Finally, thirty-four-year-old senior flight attendant José Chiu pushed hard enough, and the door blew out. Chiu was jettisoned off the airplane and killed.

On most flights the automated system works as designed. Still, at least forty to fifty times a year, an airliner somewhere in the world will encounter a rapid decompression, according to a study for the Aviation Medical Society of Australia and New Zealand. James Stabile Jr., whose company, Aeronautical Data Systems, provides oxygen-related technology, said that when slow depressurizations are figured in, the rate increases even more. And because not all events require that regulators be notified, the problem is "grossly underreported."

When planes fail to pressurize after takeoff or lose cabin altitude in flight, it is potentially life-threatening. The reason we don't see tragedies more often is because pilots are taught what to do. First, they put on their emergency oxygen masks. Then they verify that the system is on. There are numerous cases of pilots discovering that they failed to set cabin altitude upon takeoff, which I liken to finding the laundry I loaded in the washer unwashed hours later because I forgot to start the machine.

If pressurization was set correctly and is still not working, pilots immediately begin a rapid descent to an altitude where supplemental oxygen is not necessary. When pilots do not follow these steps, the situation spins out of control quickly.

To be clear, pilots don't intentionally ignore the procedures. When they do, it is usually because their mental processes are already compromised by oxygen starvation. Sometimes the effect is unfathomable; pilots faced with an alert that the cabin altitude is exceeding twelve thousand feet have been known to mistakenly *open the outflow valves*, completely depressurizing the cabin and ratcheting up the problem.

On an American Trans Air flight in 1996, a mind-boggling sequence of events brought a Boeing 727 a hairbreadth from catastrophe. The miracle is that despite the lunacy in the cockpit, the plane landed safely.

ATA Flight 406 departed Chicago's Midway Airport bound for St. Petersburg, Florida. At thirty-three thousand feet, a warning horn sounded because the altitude in the cabin was registering fourteen thousand feet. First Officer Kerry Green was flying. He immediately put on his emergency oxygen mask. Capt. Millard Doyle did not, opting to try to diagnose the problem. He instructed the flight engineer, Timothy Feiring, who was sitting behind and to his right, to silence the alarm. Doubtless already feeling the effects of altitude that was steadily increasing, Feiring could not find the control button, and more time passed.

As he looked around, the captain evidently thought he'd discovered the source of the problem, an air-conditioning pack switch that was off, and he pointed it out to Feiring. Then he turned his attention to the flight attendant in the cockpit, asking her if the passenger oxygen masks had dropped.

They had, she replied, and promptly collapsed in the doorway. Now Captain Doyle reached for his own mask, but it was too late. Disoriented and uncoordinated, he could not place it over his head, and he passed out, too.

Two of four people in the cockpit were now incapacitated, and Feiring was having trouble thinking. He mistakenly opened an outflow valve, creating a rapid and total decompression of the airplane.

He put on his mask and then got up to attend to the unconscious flight attendant, placing the flight observer mask on her face, but dislodging his own in the process. He passed out, falling over the center console between the two pilots' seats.

Through all this, First Officer Green, with his mask on, was taking the plane down to a lower altitude at a speed of about four to five thousand feet per minute.

Back in the passenger seats, the cabin crew had not been given any instructions from the cockpit, but the flight attendant seated at the front of the plane made a pantomime with her mask to demonstrate what the passengers should do. Some travelers followed her example; others did not. Through it all, the flight attendants reported that the plane was pitching up and down and side to side, and there was a brief, incomprehensible announcement from the cockpit.

Passenger Stephen Murphy of San Diego thought he was going to die and remembers feeling a sense of peace as he recited his prayers. Then the woman seated behind him started having convulsions, and the man across the aisle began to claw at his ears.

"What bothered me was there was nothing I could do for him. It's not like you see on TV; people don't grab portable oxygen bottles and walk around the cabin helping people,"

Murphy told me years later. "Had I had my full senses, I'd like to think I could have helped somebody. But based on what was going on, I didn't. I knew I couldn't."

On the flight deck, Green was trembling, a common symptom of hypoxia. Something was wrong with the microphone in his mask, and he had to pry the compressible seal away from his face to contact air traffic control.

When the oxygen mask Feiring had placed on her face rejuvenated the flight attendant, she got up and returned the favor, replacing the mask that had come off him as he moved away from the flight engineer console. She also got a mask on Captain Doyle. Soon they both came to. American Trans Air Flight 406 landed safely in Indianapolis, but the episode could have ended in catastrophe.

The story, equal parts chilling and absurd, tells me that knowing what to do does not mean pilots will actually do it if their ability to think has begun to deteriorate.

Nine years after American Trans Air 406, on August 14, 2005, a Boeing 737 took off from Cyprus on a flight to Athens, but it never arrived. Helios Flight 522 ran out of fuel and crashed into a mountain south of the airport after flying on autopilot for more than two hours—long after the pilots and nearly everyone else on board had fallen into deep and prolonged unconsciousness. They had been starved of oxygen, presumably because the pilots failed to pressurize the aircraft after takeoff. The pilots were hypoxic before they realized what had gone wrong.

The Helios 522 disaster started about five and a half minutes after takeoff, as the plane climbed through twelve thousand feet. A warning horn alerted the pilots that the altitude in the cabin had exceeded ten thousand feet.

Less than two minutes later, the passenger oxygen masks

dropped, but Capt. Hans-Jürgen Merten and First Officer Pambos Charalambous did not put on their masks, deciding instead to try to figure out what was wrong: a classic case of impaired judgment due to hypoxia.

For nearly eight minutes, Captain Merten, a pilot with five thousand hours of experience on the 737, conversed with the Helios operations center in Cyprus in an exchange that grew increasingly confusing to the men on the ground. One thing was certain. The horn warning of *altitude* did not direct the pilots to focus on the cabin altitude, and here's why: the alarm's insistent staccato is also used on the runway when an airplane is incorrectly set for takeoff. At that time in the flight, the same alarm is called a takeoff configuration warning. This case of one alarm for two hazards relies on the pilots' knowing to which hazard they are being alerted.

On the ground, it seems straightforward. The takeoff configuration alarm will sound only prior to takeoff. The distinction is not so obvious, however, when the pilot's ability to think is already fading. And we know this because, when the alarm on Helios 522 went off, Merten told his airline's operations desk that the takeoff configuration horn was sounding. He did not associate the warning with cabin altitude. That mistake has been repeated on passenger flights around the world, including ten instances over ten years found in the files of the NASA Aviation Safety Reporting System, or ASARS.

"The simplicity of the error" is what struck Bob Benzon, an accident investigator with the National Transportation Safety Board at the time, who was helping the Greeks on the Helios accident. "There were one hundred twenty-one people who died on a modern airliner, and all through a simple error. That was the thing," he said.

Six years earlier Benzon had been assigned to investigate a similar accident, involving a private jet and a popular American athlete. Payne Stewart was one of the most famous golfers on the pro circuit, beloved for the weird collection of tam-o'-shanters and knickerbockers he wore at tournaments. He suffered hypoxia on October 25, 1999, in the early stages of a four-hour flight from Florida to Texas.

Not long after departing from Orlando, the first officer, Stephanie Bellegarrigue, failed to respond to calls from air traffic control. She sounded fine in her last radio communication, but the plane failed to turn as planned, and no one on the ground could raise the crew as the plane passed thirty-two thousand feet.

"Somewhere west of Ocala, the crew became incapacitated. Maybe not dead, but they couldn't answer the radio," Benzon told me. The investigation never determined when or why the plane lost cabin pressure.

The plane continued straight from its last heading until it ran out of fuel and crashed in a field in South Dakota. From his office in Washington, DC, Benzon watched live news coverage of the runaway flight. Fifty years old at the time, he had worked nearly two hundred airplane accidents, but he had never seen one unfold before his eyes.

In the months after the Helios accident, aviation authorities in several countries shared their experiences with the investigators. Just eight months before the Helios accident, NASA had sent a special bulletin to Boeing and the FAA, concerned that several flight crews reported they had been confused by the dual use of the pressurization warning horn. Even earlier, in 2001, there had been an event in Norway when pilots disregarded the warning horn and continued to ascend. The Norwegian Air Accident Investigation Board sent a safety recommendation to

Boeing also in 2004, calling for it to discontinue the dual-use alarm.

As Helios 522 ascended over Cyprus, Captain Merten's thoughts were scattering, and his brain was going dim. He collapsed at his last position, checking a switch panel behind his seat. First Officer Charalambous passed out against the airplane control yoke.

Using the experience of the survivors of American Trans Air Flight 406 as a reference, we can assume that the passengers on the Helios 737 were uneasy once their masks dropped, everyone waiting for news from the flight deck. But that uneasiness would not have lasted for more than twelve to fifteen minutes, because those masks have only a limited supply of oxygen; after that, the passengers would have passed out. This is why pilots quickly have to get the airplane to a lower altitude, but the pilots on Helios 522 were unconscious, and they weren't going to recover. There was no one to initiate a descent, and the plane flew on, northwest past southern Turkey and high above the Greek islands.

The flight attendants had higher-capacity emergency oxygen bottles and portable oxygen masks. With more than an hour's supply in each, they were likely conscious longer than the passengers. Twenty-five-year-old Andreas Prodromou was a flight attendant who also happened to be a private pilot. He may have waited for word from the cockpit, but at some point he got up from his seat by the back galley and took action.

What we know from this point comes from two sources: recordings in the cockpit documenting Prodromou's arrival on the flight deck and the observations of two Greek fighter pilots who were dispatched to see what was happening with the airliner that had silently, and without contacting controllers, entered Greek air space.

Two air force F-16s were flying on either side of the airliner. It was just four years after terrorists had crashed four commercial jets into landmarks in New York and Washington, DC, and the Greek Air Force pilots expected to find something similar. Instead, they saw the first officer unconscious in the right-hand seat. One of the air force fliers saw Prodromou enter the cockpit. This means that Prodromou waited more than two hours after the depressurization.

He may have suspected the incapacitation of the crew, but the sight of the vacant captain's seat and the copilot lifeless at the controls must have been terrifying. Captain Merten was partially on the floor and partially on the center console. Prodromou probably had to step over him to get to the left seat, where he removed Merten's unused oxygen mask from the storage compartment and put it on. Lifting the mask activates the flow of oxygen through a thick gray umbilical cord that also contracts the face straps. This design keeps the mask fitted tightly to the head.

Prodromou put on the mask as the last of the left engine's fuel was spraying into the combustion chamber. In moments, the engine would stop producing power.

Bank angle, bank angle. A computerized voice warned that the airplane's left wing was losing lift. The Boeing 737 can fly with only one engine, but control surfaces have to be adjusted to compensate.

Prodromou searched the control panel for something familiar—something that connected this complicated aircraft to the small planes on which he had learned to fly. Then the control wheel in front of him started to vibrate. The stick shaker warning is as dramatic as it is urgent, an attention-getting, multisensory advisory that the plane is about to stall. For two and a half minutes Prodromou scanned the instrument panel while the airplane picked up speed in descent. The sound of rushing air joined the cacophony of warnings. Finally, hope extinguished, he called for help in a frail and frightened voice.

"Mayday, mayday, Helios Flight 522 Athens . . ."

And forty-eight seconds later:

"Mayday."

"Mayday."

Traffic, traffic. He heard only the mechanized voice of the 737.

The radio was not set to the proper frequency to transmit the message. Prodromou's mayday would be heard only in the postcrash examination of the cockpit voice recorder.

As the plane approached the ground and ambient air pressure increased, the cabin altitude warning horn turned off and one contributor to the din in the cockpit subsided. It was then that Prodromou first noticed the fighter jet escort.

Years later, one of the fighter pilots explained that he gestured for Prodromou to follow him to a military airfield nearby. To this signal, the young flight attendant raised his own hand and, with weary resignation, pointed downward. Even if he could have figured out how to follow the F-16, it was too late. The right engine was shutting down. The plane was seven thousand feet above the ground with three and a half minutes left. Helios Flight 522 crashed into the countryside near Athens International Airport, not far from where it had been programmed to fly, killing the last of the travelers on this terrible journey.

When Prodromou's role in the story made the news in Cyprus, many wondered: what if the young man had entered the cockpit earlier? Many factors could have changed the course of Flight

522. But at its heart, what claimed Prodromou and the others was a simple truth.

"Inherent in aviation is the exposure to altitude," said Robert Garner, an aviation physiologist and director of a high-altitude training chamber in Arizona, "and the risk of hypoxia is always present."

Emergency

In the early days of the Malaysia 370 mystery, I thought of these episodes. After all, it was an ordinary flight—under the command of an experienced and well-regarded captain—that suddenly turned baffling.

The Boeing 777 departed Kuala Lumpur International Airport on March 8, 2014, on an overnight trip to Beijing. There were 227 passengers and 12 crew members on board. In the cockpit, Capt. Zaharie Ahmad Shah, a thirty-three-year employee of the company, was in command. He had eighteen thousand flight hours. As a point of reference, that's just fifteen hundred hours fewer than Chesley Sullenberger had in his logbook when he successfully ditched a disabled US Airways airliner into New York's Hudson River, and Zaharie was five years younger than Sully.

Zaharie spent even more untallied time flying his home-built flight simulator. He took so much pleasure in this activity that he made videos and posted them on his Facebook page, offering tips and instructions to other simulator enthusiasts. *Obsessed much?* you might think when I tell you he also owned and flew radio-controlled airplanes. There just wasn't enough flying, as far as Zaharie was concerned.

Professionally speaking, First Officer Fariq Abdul Hamid was everything Zaharie was not. Inexperienced on the Boeing 777, he was still training on the wide-body while Zaharie supervised his performance. The flight to Beijing would bring the young pilot's total hours on the airplane to thirty-nine. Fariq had been flying for Malaysia for four years. From 2010 to 2012, he was a copilot on Boeing 737s. He was moved to the Airbus A330, where he flew as a first officer for fifteen months until he began his transition to the even bigger Boeing 777.

The moonless night was warm and dark with mostly cloudy skies when the jetliner lifted off at 12:41 a.m. on Saturday morning. Fariq was making the radio calls, so we can assume Zaharie was flying the plane.

On board were business travelers, vacationers, and students. There were families, couples, and singles from Indonesia, Malaysia, China, Australia, America, and nine other countries; a global community common on international flights. Because Kuala Lumpur and Beijing are in the same time zone and the flight was to arrive at dawn, many travelers were probably sleeping when things started to go wrong.

Flight 370 was headed north-northwest to Beijing. Twenty minutes after takeoff, at 1:01 a.m., the plane reached its assigned altitude, thirty-five thousand feet, and Fariq notified controllers.

"Malaysia Three Seven Zero maintaining flight level three five zero."

Independent of what the pilots were doing, the twelve-year-old Boeing 777 was transmitting a routine status message via satellite with information about its current state of health. In the acronym-loving world of aviation, this data uplink is called ACARS, for Aircraft Communications Addressing and Reporting System. Messaging can be manual if the pilots want to request or send information to the airline. It can also be triggered by some novel condition on the plane requiring immediate notice. Absent either of these conditions, an automatic status report is transmitted on a schedule set by the airline. At Malaysia, it was every thirty minutes.

Pilots may not be aware of when or how often the aircraft makes scheduled status transmissions, but they certainly know about them. They use ACARS often, for both the serious and the mundane things that happen in flight, from requests for weather updates to the latest sports scores. A pilot who needs a minor repair or a wheelchair on arrival can simply send a text through ACARS.

Neither Zaharie nor Fariq had anything to add to the 1:07 a.m. scheduled report, and the message showed nothing amiss. The engine performance indicated how much fuel had been consumed by the Rolls-Royce Trent 892 engines.

Around the time the ACARS message was being sent, it appears control of the flight was transferred to the first officer because Captain Zaharie was now making the radio calls. He confirmed to air traffic control that the plane was flying at cruise altitude. "Ehhh . . . Seven Three Seven Zero* maintaining level three five zero."

Eleven minutes later, as the airplane neared the end of Malaysian airspace, the controller issued a last instruction to the men in

* This was an error, as the flight was Three Seven Zero, not Seven Three Seven Zero.

command of Flight 370, giving them the frequency to which they should tune their radio upon crossing into Vietnam's area.

"Malaysian Three Seven Zero contact Ho Chi Minh one two zero decimal niner, good night."

"Good night, Malaysian," Zaharie said. It was 1:19. His voice was calm, according to a stress analyst who listened to the recording as part of the Malaysian probe. There was no indication of trouble.

Zaharie, fifty-three, had been in his seat since around 11:00 p.m., ordering fuel, entering information in the onboard computers, arming systems, checking the weather en route, and discussing the flight with the cabin attendants. He had also been supervising Fariq, who, after landing in Beijing, would be checked out on the Boeing 777. That was sure to be a heady and exhilarating new assignment for the young man, as Zaharie certainly recognized, having three children of his own around the age of Fariq, who was twenty-seven.

The airliner was at cruise altitude, flying a preprogrammed course. There was very little difference at this point between the Boeing 777 and every other jetliner Fariq had flown. So, in the scenario I envision on Malaysia 370, this would have been the perfect time for Zaharie to tell Fariq, "Your airplane," leaving the triple seven in the first officer's hands so he could go to the bathroom. And so he did.

Alone on the flight deck, Fariq must have enjoyed these moments. He was in sole command of one of the world's largest airliners, responsible for taking his passengers to their destination.

Seven years earlier, he had graduated from junior science college, a boarding school three hours north of his family home in Kuala Lumpur. He was accepted into Malaysia Airline's pilot cadet program, at the Langkawi Aerospace Training Centre, on the northwest coast of the Malay Peninsula. He would get more than flying lessons at the training center. He had a guaranteed job flying for his nation's flag carrier, which served sixty destinations around the globe and operated the Airbus A380, the world's largest airliner.

His professional future was full of promise and so was his personal life. During cadet training he met and fell in love with a fellow student, Nadira Ramli, who became a first officer with AirAsia, a Kuala Lumpur–based low-cost carrier. Ramli, one year younger than Fariq, was so charming that she was selected by AirAsia to represent the company on a public relations and marketing campaign that included a drive across China in 2012. In March 2014, Fariq and Ramli were engaged to be married.

While Zaharie was out of the cockpit, it would be Fariq's job to tune the radio to the Ho Chi Minh air traffic control frequency. Once he established contact, he would change the transponder's four-digit squawk code from the one used in Malaysia to one for transiting to Vietnam-controlled airspace. But instead of making that switch, the transponder stopped transmitting entirely. The question is why.

A transponder is critical for airliners. It links altitude, direction, speed, and, most significantly, identity to what otherwise would be a tiny anonymous green dot on an air traffic control screen. The transponder provides what is called a secondary return: a data-rich reply to a radar interrogation. Controllers need the transponder to keep planes from colliding in increasingly crowded skies. Airlines use it to track the progress of flights. Pilots depend on it for a timely warning if another plane winds up in their flight path.

Turning the knob on the lower-right-hand side of the device to the left—to the "standby" mode—effectively shuts off the transponder. This stops sending the plane's identification information and eliminates the plane's ability to be seen on the collision avoidance systems of other aircraft. Standby is used mostly while airliners are taxiing at the airport, so all the planes don't trigger the collision avoidance system. For all intents and purposes, standby is "off."

In flight, however, pilots have little reason to turn off the transponder, though there have been cases in which it has stopped transmitting for undetermined reasons. In one perplexing flight in 2006, the lack of secondary radar contributed to a catastrophic collision. An Embraer Legacy business jet was being delivered to New York from Manaus in Brazil. While flying over a remote jungle region, it hit a GOL Airlines Boeing 737 flying in the opposite direction at the same altitude. The pilots of the Legacy said they did not intentionally shut off the transponder, but it was in standby mode, so as they flew westward at thirty-seven thousand feet, an altitude normally reserved for eastbound flights,* the plane was invisible to the GOL Airlines collision avoidance system. These systems require both planes to have a transponder operating to issue an alert.

The winglet of the small business jet sliced through the left wing of the 737, and the airliner fell from the sky and into the jungle, killing all 154 people on board. The little jet made an emergency landing, and no one aboard was injured. When

* Commonly, eastbound flights fly at odd-numbered altitudes, and westbound flights at even-numbered altitudes.

questioned, neither pilot on the business jet was able to explain what had happened.

On the captain's side of the Embraer, the switches for the Honeywell transponder are positioned below a bar that is also used as a pilot footrest. Investigators suspected the captain might have kicked the switch into standby mode. There was also a theory that the lid of the laptop both pilots had been using might have pushed the button into standby. Less than a year earlier, Honeywell discovered a software glitch on more than thirteen hundred devices that could make them go into standby mode if pilots failed to dial the squawk code in less than five seconds. So there were plenty of theories about what might have happened. Ultimately, though, the Brazilians concluded that the software problem was not an issue on the Legacy involved in the collision, and what happened in that case remained a mystery.

In the United States, however, the Brazilian accident led the safety authorities to one unequivocal conclusion: something needed to change. The FAA issued a rule in 2010 that on new airplanes, the warning of an inoperative transponder should be more obvious to the pilots. Planes produced before the rule, including 9M-MRO, the registration for the plane that was flying as Malaysia 370, would not be affected.

Zaharie left the cockpit for what is delicately called a "biological break." Perhaps he would have stopped by the galley for a cup of coffee or a snack. It's a long flight at cruise altitude, so there would have been no rush to get back to the flight deck. The tasks that First Officer Fariq had to take care of were routine. Easy peasy, as they say.

Fariq knew he had to get the squawk code from Ho Chi

Minh—but first he had to tune the radio to that frequency. This is about the time when, I think, a rapid decompression happened near or in the cockpit. It would have made a deep and startling noise, like a clap or the sound of a champagne bottle uncorking, only much, much louder and sharper. This would have been followed by a rush of air and things swirling everywhere. Zaharie's nearly empty coffee cup, pens, papers—everything loose—would have been tossed about in the wind, including the shoulder straps of Fariq's seat restraint, which he would have unfastened for comfort not long after the plane's wheels left the runway at Kuala Lumpur. A white fog would have filled the space as the drop in temperature turned the moist cabin air into mist. The first officer would have realized immediately, *This is an emergency.* It would have been a neon light in his brain, but it would also have been competing with other lights and sounds, physiological sensations that had to have been both disconcerting and overwhelming.

The denser air inside Fariq's body would have rushed out through every orifice, an effect that can be particularly painful in the ears, as anyone who has flown with a head cold already knows. His fingers, hands, and arms would have started to move spastically. Fariq would have struggled to understand this rapid change from normal to pandemonium while irretrievable seconds of intellectual capacity ticked away.

Emergency, have to get down, have to let someone know. What first? He would have reached over to the transponder to enter 7700, the four digits that will alert everyone on the ground and in the air that something has gone wrong with the plane. His fingers would still have been trembling as he clutched the small round knob on the bottom left of the device and turned it to Standby. It is not what he would have intended, but he would already have begun to lose his mental edge. In an attempt to transmit a message of distress, he would have inadvertently severed the only means air controllers had of identifying his airplane and the details of his flight. It was half a minute past 1:20 in the morning.

A Fading Glimmer

It is not difficult to imagine Fariq responding inappropriately. As the Greeks investigating the Helios disaster discovered, only a small portion of pilots has experienced the dangerous seduction of hypoxia. Military aviators in many countries are trained to recognize the symptoms of oxygen deprivation by spending time at twenty-five thousand feet in high-altitude chambers. Yet even military pilots, astronauts, and soldiers are not subjected to the kind of rapid decompression that could have happened on MH-370. The onset of hypoxia above twenty-five thousand feet is too quick, and the health risks too high, to duplicate it in a high-altitude chamber.

When MH-370 lost secondary radar and disappeared from controllers' screens at thirty-five thousand feet, the plane wasn't exactly invisible. A two-hundred-foot blob of metal can hardly be missed by the sweep of a radar signal, even if the antenna is as far as two hundred miles away. However, the signal sent back, called an echo, does not transmit the precise information provided by the transponder. The object is picked up on the radar sweep in what is called "primary" mode. Things as different in size and nature as a flock of geese, a cloud, or a ship can all cause the radar signal to ricochet, and show up on the screen as a green blip.

Primary radar is a "no-frills" target. Viewing these kinds of blips over time allows calculation of an object's speed, which can help determine if it is an airplane, as few things move as fast as an airliner. Sometimes it is possible to tell what kind of plane it is because different planes move at different speeds. The Boeing 777 cruises at around five hundred seventy-five miles per hour.

Altitude is a different story. It is much more complicated to judge height, and altitude cannot be determined from primary civilian radar. Only military radar has this capability.

After MH-370 went missing, stories and theories emerged based on a Reuters wire service report that the airliner went on a wild ride of ascents and descents after turning back toward Kuala Lumpur. While this was based on real information collected and reviewed by international military and civilian radar specialists, some of the data was "essentially useless," according to one of the men who participated in the evaluation and who wishes to remain anonymous.

Not all the air defense systems capable of capturing altitude actually got it, and among the altitude data collected were indications that the target thought to be the plane was dropping thousands of feet in a few seconds. This had to be considered erroneous, because the plane could not move that quickly.

"It was being reported accurately as far as it went. It was showing a forty-thousand- to twenty-five-thousand-foot descent, but to make an airliner do that would require a ten-thousand-foot-per-minute descent," I was told by my source, more than twice even a rapid ascent rate. "A lot of the numbers were not reasonable."

What was never reported is that this questionable altitude information caused a controversy among those reviewing the tapes, because some civilian radar specialists thought it indicated that the plane had been hit by a missile. This dominated the discussion for several days, with the Malaysian Air Force arguing against the theory. What settled it, according to the participant who told me about it, was the lack of wreckage in the South China Sea. "The search was going on in that area, the last place the airplane was seen, but they weren't finding anything," this person said. "If it had been shot down you would have found pieces of stuff, but there was no evidence to back up that theory, so we came to a consensus that's not what happened."

That consensus got another boost when the engineers from the satellite company Inmarsat showed up in Kuala Lumpur a few days later to share with the team information that the airplane had not come to a sudden end after disappearing from radar, but flew on a lengthier and far more puzzling journey. They knew this because the airplane was exchanging digital handshakes with a communications satellite. The logs of those exchanges also provided a small slice of data about 9M-MRO's final hours.

Before the plane departed Kuala Lumpur it was loaded with just under 111,000 pounds of jet fuel. Based on fuel consumption between 15,000 and 17,000 pounds per hour on a Boeing 777-200, at best the plane had 7.2 hours of flying time. The Inmarsat data showed the plane did fly slightly longer, for 7.5 hours, meaning it could not have engaged in steep ascents or low-altitude flying, both of which burn more fuel.

The satellite data also indicated that nothing could have happened to the airplane to cause a decrease in its performance, such as a debilitating fire or structural damage. These would have caused more drag or prevented the plane from remaining aloft as long as it did.

Like the no-frills radar data, the equally unpretentious signals between the airliner and the satellite communication network would become a significant source of information, providing facts even the experts didn't know they had.

While the radar intermittently picked up the presence of something moving at the speed of a 777 heading southwest over the peninsula, inside the cockpit of 9M-MRO, Fariq's brain would have been hovering in a state of befuddlement. He would have been not in the game but not entirely out of it, either. When the interior atmosphere of the 777 suddenly zoomed from eight thousand to more than thirty thousand feet, the young pilot did the wrong things as his rapidly diminishing mental state was telling him he was doing the right things. He would not have become aware of his error: hypoxia victims think they are performing brilliantly.

When I try to imagine Fariq's compromised intellectual state, I recall an army aviator in an altitude chamber training session, later posted on YouTube. I could not stop watching the astonishing transformation of the man in the video, identified as Number 14.

The young soldier is flanked by two others using supplemental oxygen, but Number 14 has his regulator off in order to experience hypoxia. He holds a deck of cards and has been asked to flip through them one by one, announcing the number and suit before moving on to the next card. The altitude in the chamber is twenty-five thousand feet.

"I feel really good right now," Number 14 says as he begins announcing, "Six of spades," and showing a six of spades to the camera. "No symptoms yet." In twenty-four seconds he reports feeling tingling "in my toes and in my toes." One minute in, Number 14 gets his first card wrong. He identifies a five of spades as a four of spades. After being asked twice to look again and making the correction, he calls every card the four of spades.

After two minutes, as his thinking gets increasingly sloppy, Number 14 is asked, "Sir, what would you do if this was an aircraft?" to which he replies, "Four of spades, four of spades." Ninety seconds later, after ignoring several requests that he put on his regulator, a seatmate does it for him.

Sessions like this are intended to demonstrate to future pilots the danger of hypoxia. Like the drunk who's convinced he's the funniest guy in the room, a pilot suffering from hypoxia can feel a heightened sense of competence and well-being, what one pilot called "a lightheaded euphoria."

This is a tricky issue, because hypoxia can lead to brain death. People experiencing it ought to be trying to get some oxygen STAT, but they often don't. Hypoxia creates a state of idiotic bliss. One commenter on the YouTube video wrote, "Make this legal," because it sure does look like silly fun.

I expected to have a similar experience when I joined two dozen pilot cadets from Taiwan's EVA Air for a daylong hypoxia awareness training session at the Del E. Webb High Altitude Training Chamber at Arizona State University Polytechnic at Mesa, run by hypoxia specialist Dr. Robert Garner.

Prepared for the goofy, loopy, playful effects of oxygen deprivation, I removed my mask when the hypobaric chamber reached the atmospheric equivalent of twenty-five thousand feet. I began carefully doing the simple math problems on the clipboard given to me and smugly noted that I was getting them right. Some of my classmates were also diligently writing, but others were looking around grinning. My fellow student Shih-Chieh Lu said the sensation was like that of being drunk. After about one minute, my breathing was labored. The head lolling started two minutes in.

"Hot," I said, more an exhalation than a statement, because it required a lot of effort just to push the microphone button to speak to the chamber operator. That was it. I passed out, and chamber attendant Dillon Fielitz got an oxygen mask on my uncooperative head. In another minute, the oxygen worked its magic and I was roused from my oblivion. I was unaware that I had lost consciousness and had no recollection that Fielitz had come to my aid.

"This high-altitude chamber training experience is quite helpful to the pilot," another cadet, Yuchuan Chen, told me in an e-mail later. "It will become a hazardous situation if pilots encounter the loss of pressurization without any correction." Silliness or unconsciousness, the symptoms can vary, but Yuchuan's impression is the hoped-for takeaway of altitude chamber training, Dr. Garner said.

Hypoxia was responsible for at least seven fatal aviation accidents since 1999, and many more near disasters. In 2008, both pilots on a Kalitta Flying Service Learjet became hypoxic at thirty-two thousand feet. The flight had just been handed over to Cleveland ATC (air traffic control) when the controller became concerned about the halting transmission of the pilot and the sound of an alarm in the background. In what seems like farcical overenunciation that makes a lot of sense to me now, the captain explained to the controller that he was "Unable . . . to . . . control altitude. Unable . . . to . . . control . . . airspeed. Unable . . . to . . . control heading." He added, "Other than that, everything . . . A-OK."

It must have taken extraordinary effort for the Kalitta pilot to pierce his mental fog enough to make the emergency call. Recognizing the problem, controllers cleared the Learjet to an immediate descent to eleven thousand feet. What made the pilot perceive and react to the instruction remains a mystery and a miracle, as it certainly saved lives. The plane flew lower, the crew revived, and the plane landed safely.

A similar scenario on Malaysia 370 doesn't explain everything, but it does explain a lot. Fariq would have known right away that he had a problem, even without the steady high/low electronic sound of the altitude warning horn. And at some point he must have remembered to put on his oxygen mask. It was stored in a chamber the size of a car glove compartment, below his armrest. He may have been slowed by his sluggish movements or confounded by the difficulty of squeezing the red tabs together with his thumb and middle finger so that the huge circle of rubber would expand enough to go over his head before he released the tabs to shrink it back to secure the regulator tightly over his nose and mouth. Thick clear plastic goggles covered his eyes, and a gasket should have created a seal.

So what happened? Why did it fail to revive him? Why didn't Captain Zaharie return to the cockpit? Everything was in chaos, the altitude warning alarm still clanging.

I find it logical to assume that Zaharie visited the business-class bathroom near the flight deck that is also used by the flight crew. In this and all the airline's 777s' bathrooms, a drop-down mask is there to provide oxygen in the case of depressurization. Imagine what it would have been like for Zaharie to see the yellow plastic cup bob down after the depressurization. He would have been momentary rattled, but with his experience, he would have realized immediately what had happened and what needed to be done.

Still, he had to make a choice: try to get back to the cockpit without supplemental oxygen, or remain in the bathroom and wait for Fariq to get the airplane to a lower altitude and then rejoin him on the flight deck. I'm guessing Zaharie wasn't confident in Fariq's ability to handle the emergency and chose the former course of action. But the effect of oxygen deprivation would have been crippling for Zaharie, too. Air would have been exploding from his respiratory and digestive systems. His extremities would have been shaking. He would have struggled to get out of the bathroom. Perhaps he looked for a flight attendant or a portable oxygen tank. Perhaps he stopped to assess the situation in the cabin. Perhaps he retained focus and moved quickly to the cockpit door.

The distance between the bathroom and the cockpit is just a few steps, but like Fariq, Zaharie was a smoker and probably more susceptible to the effects of oxygen deprivation. If he got out of the bathroom, if he got down the narrow corridor, if he got to the door of the cockpit without losing consciousness or cognitive function, another challenge would have awaited him.

The cockpit door unlocks automatically when cabin altitude is lost. Would Zaharie have remembered that? Or did he, by force of habit, stop outside the door and try to enter the code? Did he lose precious seconds struggling to remember a passcode he did not need? Or did he just grab the handle and open the door, but succumb to the lack of oxygen before getting into his seat? Pilots at Malaysia Airlines tell me that in a rapid decompression, it would have been very difficult for Captain Zaharie to get back onto the flight deck. All the previous cases of rapid depressurization on airliners, those that successfully landed and the few that crashed, bring home with chilling clarity that physical exertion eats away at the too-few seconds of useful consciousness.

The captain was unable to regain command of the airplane. If he had, things might have turned out very differently.

Incomprehensible

On the flight deck, Fariq was wearing his oxygen mask. He was getting enough oxygen to sustain some level of intelligent thought. *Why hasn't Captain Zaharie returned?* he must have thought. And he must have realized that he needed to get the plane back on the ground.

The control panel for the flight management system is located between the two pilot seats, above the throttles, where it is easily accessible to whichever pilot is programming it. The FMS has many functions, including allowing the crew to send text messages to the airline's operations desk. We know no messages were sent. Yet in an emergency, the FMS stores navigational information for the closest airports, so that in seconds the pilots can select a destination and head there.

From where the 777 was flying, between the Gulf of Thailand and the South China Sea, if Fariq turned the plane around, the divert airports would include Penang and Langkawi, according to pilots who fly in the region. These choices would have appeared on the screen in a list, waiting for the pilot to select one of them.

Who knows how much actual thinking Fariq was able to accomplish, but for some reason he selected Penang, Malaysia's third-busiest airport, with a ten-thousand-foot runway. The next choice appeared on the screen. DIVERT NOW? Fariq selected, EXECUTE.

The plane immediately began a slow, orchestrated turn, and by 1:30 it was headed south-southwest to Malaysia, once again.

The amount of time a person can remain conscious and thinking at high altitudes is called the time of useful consciousness. While that time varies depending on many factors, including health, age, and a genetic predisposition, the ballpark figure for how long Fariq had before he lost his ability to think clearly would be fifteen to thirty seconds. We know that Fariq, or whichever pilot was in the cockpit, maintained sufficient intellectual capacity to turn the airplane around and select a course toward Penang. Yet that these maneuvers were made without a radio call and after the transponder became inoperative leads me to conclude that the pilot handling the airplane was compromised to such an extent that while he could make simple decisions about the direction of the airplane, not much more sensible action could have been expected of him.

Fariq was breathing through a mask. The default position should have given him 100 percent oxygen, and at thirty-five thousand feet, positive pressure actually pushes the oxygen into the wearer's nose. I experienced this during my time at the ASU high-altitude training chamber. It felt like an air-conditioning vent was being pressed to my face.

When all is working well, the mask should rejuvenate. Fariq's vision would have been clear again and his thoughts solidified, except, judging from what happened next, he did not return to his senses. The primary indicator of that is that the plane did not start to descend.

Because of the seriousness of loss of pressurization in flight, the modern airliner has a belt-and-suspenders approach to the hazard. The oxygen mask is the belt, and emergency descent is the suspenders. They are equally important, two routes to the same destination: clearheadedness.

In his book *Of Flight and Life*, Charles Lindbergh tells of testing an unpressurized fighter plane at thirty-six thousand feet in 1943 when his oxygen supply abruptly stopped. "I know from altitude-chamber experience that I have about 15 seconds of consciousness left at this altitude—neither time nor clearness of mind to check hoses and connections. Life demands oxygen and the only sure supply lies four miles beneath me," he writes.

As he recounts in the book, Lindbergh sent the airplane into a dive, rocketing toward earth as he passed out. Not until he was at fifteen thousand feet did he come to and witness the clarity of "the cockpit, the plane, the earth and sky."

That was not Fariq's experience. The razor's edge was dull. His mask was providing him with enough oxygen to maintain some awareness, but he was not thinking clearly.

Any number of problems may have prevented Fariq from getting enough oxygen even while wearing his mask. Something wrong with the mask, the oxygen supply, or the connection between the two could explain why he might not have experienced what Lindbergh called "the flood of perception through nerve and tissue."

In the hours before MH-370 departed for Beijing, mechanics had serviced the two oxygen containers for the cockpit, topping them off and restoring the pressure to eighteen hundred pounds per square inch. After reinstalling the bottles, mechanics must reopen the valve fully, or the proper supply of air will not flow to the mask. "One or two times a year out of the hundreds of times oxygen bottles are changed at a major U.S. airline, a mechanic may fail to do this," according to a mechanic who agreed to discuss the issue if I did not identify him.

"It's a lapse in memory, and it's embarrassing," he told me. I was asking my contact about this because of a story I heard from a pilot who flies for a different U.S. carrier. The pilot was conducting his preflight check when he discovered there was little oxygen flowing to his mask. "I had the mechanic come to the cockpit," he told me—again, as long as I did not use his name—and it was then that they discovered the supply valve was barely open. "He was shocked; he was ashen," the captain said, describing the mechanic, who then got a little spooky. "You all would have died," the mechanic told him.

On many airliners, this important final action after servicing the oxygen is not left to a mechanic's memory. A message appears on a flight deck monitor notifying the pilots if the oxygen pressure drops between the tank and the mask. If the supply line between the tank and the mask is full, the indicator will show that the oxygen system is working properly, but it does not indicate if the valve is only partially open, which would reduce the oxygen available to the pilot in a depressurization event.

"To the pilot doing the preflight, it looks, because of the trapped air in the supply line, that the system it is fully pressurized, and if he looks at the monitor it will show the tank is fully charged," this pilot told me. If the crew needs oxygen during the flight, that restricted flow could cause a problem for the pilots. "Once that stored volume of oxygen in the supply line flows out, the pressure will drop within this line to some value that is insufficient. It won't supply full oxygen, so no matter how hard he breathes, he is not going to get enough oxygen."

There are other potential pitfalls. Leaks in the supply tubes or in the seal holding the mask on the face can diminish the supply to the pilot. While working for the NTSB, Dr. Mitch Garber said he would sometimes fly in the cockpit observer seat. On three occasions he discovered a problem with his oxygen mask. Once, the air-filled tubes that contract to hold the mask to the head were leaking. Another time, the inflation of the tubes was followed by a loud pop and no air flow. The one he remembers best is the time the mask worked fine in the box, but when he pulled it out of the holder, it fell apart. "That was the one that got me kicked off the plane, because there were no other masks," he said, adding, "These things sit in these boxes for a long time."

Another factor that could have kept Fariq from regaining full cognitive function was if the aneroid barometer in the regulator of his mask failed so that it did not correctly sense cabin altitude. Above thirty-five thousand feet, this small bellows-like device triggers the mask to provide not just a mix of pressurized air but 100 percent oxygen under pressure.

In a decompression at higher altitudes there is a delay between a pilot's first breath of supplemental oxygen and its arrival in the brain. James Stabile, an airline pilot and a longtime member of the industry committee overseeing standards for aircraft oxygen systems, asked me to imagine little boxcars loaded with oxygen, chugging first from the lungs, where oxygen will enter the bloodstream, then to the heart and then to the brain.

When the oxygen pressure drops suddenly, as in a rapid decompression, gas races out of the body, including out of the lungs. The time it takes for this oxygen shortage to reach the brain is about ten to twelve seconds. That's the time of useful consciousness at high altitudes. Pushing 100 percent oxygen into the lungs will enable the next several boxcars to resupply the brain, switching it back on, and "quite often the individual will not even be aware of this cognitive lapse," Stabile said.

The difference between what happened on Helios Flight 522 and the private jet carrying Payne Stewart and what I believe occurred on 9M-MRO (the plane that was Malaysia Flight 370) is that when it departed Kuala Lumpur, the cabin was pressurized. Had it not been, the pilots' exchange with air traffic control at 1:19 a.m. would have indicated that something was amiss. The problem would also have been transmitted via the 1:07 a.m. normal ACARS status report. What happened on Flight 370 happened suddenly.

Because pilots succumb to hypoxia so quickly at cruise altitude, some government aviation regulators require that if one pilot leaves the cockpit, the crew member remaining wear the emergency oxygen mask. And while the intent is good, the execution is inconsistent. Crews frequently ignore the rule. One pilot told me he had not put on his oxygen mask in five years; nor had he been asked to by a fellow pilot vacating the cockpit. "It is incredibly cumbersome," he told me.

John Gadzinski, a pilot with a U.S. airline and a private safety consultant, told me why so few pilots comply. "You have to take off your headset and put it back on and maybe even take your glasses off. You then have to speak through the microphone in the mask and reset the communications when you stow the mask again," he told me. "Pilots are human, and ninety-nine-point-nine percent of the time, nothing bad ever happens on a flight."

So I think that when Zaharie left the cockpit, leaving Fariq at the controls, the young first officer did not put on his mask. Neither pilot anticipated the number of things that could have gone wrong, from the banal to the bizarre. Here are a few:

In 2011 a rupture in the roof of a Southwest Airlines Boeing 737 at thirty-four thousand feet caused a rapid loss of pressure on a flight in Arizona. Passenger masks dropped, but one flight attendant who was trying to use the public address system before putting on his mask lost consciousness, as did the passenger who tried to help him. The pilots made an emergency descent and landed without further problems.

Faulty door seals and breaks in the structure of an airplane have been known to cause decompressions. In one case, a passenger oxygen bottle exploded on a Qantas Boeing 747 in 2008. The bottle shot through the side of the airplane like a small missile, leaving a hole large enough to cause a rapid decompression. No one was injured.

Sometimes, however, decompressions do turn deadly. In one horrific case in 1988, an eighteen-foot section of an Aloha Airlines 737 tore off on a flight to Honolulu, sucking a flight attendant out of the airplane.

British Airways Flight 5390 is another macabre story. This was an early morning trip from Birmingham, England, to Málaga, Spain, on a sunny day in June 1990. As the BAC-111 jet with eighty-one people on board passed through seventeen thousand feet, the cockpit windshield blew out. Capt. Tim Lancaster was partially sucked out of the hole, but his legs got tangled in the flight controls.

Flight steward Nigel Ogden had just turned to leave the flight deck, after checking to see if the pilots wanted tea, when he heard the blast. He thought a bomb had gone off. When he turned around, he saw the captain's legs.

"I jumped over the control column and grabbed him round his waist to avoid him going out completely," Ogden wrote in a first-person account for a local newspaper.

In his unexpected exit through the cockpit window, Lancaster had kicked off the autopilot. While another flight attendant raced in to help keep the captain from disappearing, First Officer Alistair Atcheson regained control of the airplane, and then prepared for an emergency landing, which he accomplished just eighteen minutes later. Captain Lancaster survived, and returned to flying. An examination of the airplane showed that while replacing the windshield days earlier, a mechanic had used screws slightly shorter than those required, so the new window was not effectively secured.

So you can see that in the case of Malaysia 370, a loss of pressurization mishandled by the pilot is neither farfetched nor unprecedented. It fits the facts we know.

By 1:52 a.m., Fariq had taken the plane back across Malaysia and to Penang. Here he made yet another decision explicable only by a hypoxia-induced, half-witted state. He turned the plane north. Perhaps he had the intention of landing at Langkawi International Airport, where he'd learned to fly. Surely the airfield was as familiar to him as his own driveway, and the runway was nearly two thousand feet longer than that at Penang. He would be coming in heavy, with much of the fuel loaded on the plane in Kuala Lumpur still in the tanks. If Fariq did any mental processing at all, he may have concluded the more runway, the better, and Langkawi had a lot of it. Yet I think he was no longer doing much reasoning, because his ability to do that was long gone. Turning to the northwest, 9M-MRO continued to fly. There was no effort to descend or to begin an approach to the airport. Fariq had been flying for thirty-two minutes since the occurrence of whatever had caused the flight to go amiss. Still at cruise altitude, the plane passed over VAMPI—one of the many navigational waypoints in the sky, all of which have five letter names. Then the plane flew north of the next one, MEKAR, disappearing for good somewhere at the northernmost part of Sumatra.

Intermittent Power

Fariq's mental incapacitation explains a series of perplexing events that began with a sudden and unknown catastrophic occurrence. Some have theorized it was related to the load of lithium-ion batteries the plane was carrying. That's an iffy theory to me, for two reasons. First, a lithium-ion battery fire is a frightening thing, which you will read about later in this book. I have little doubt that in such an alarming circumstance, the pilots would have understood the need to get the airplane on the ground quickly. Moreover, had there been a fire, it is unlikely it would have disabled the crew without causing significant damage to the structure of the airplane, and we know the plane continued to fly with notable efficiency for many more hours after the initial problem.

Whatever happened to Flight 370 probably caused both the depressurization and an encompassing failure in the airplane's electrical system. It is not knowable if Fariq accidentally turned off the transponder or if it failed on its own. The same is true for the loss of the ACARS reporting system: did it fail or was it intentionally switched off for some reason? Yet an even more intriguing clue is the loss of regular transmissions from the plane to the satellite sometime between 1:07 a.m. and 1:37 a.m., with the return of the signal at 2:25 a.m. Even those paying attention to the Flight 370 story have heard little about this peculiar lapse.

During the ongoing news coverage, people learned that airliners regularly transmit a status message: a "ping" or "handshake" in the same way a mobile phone that is powered on sends out signals to nearby cell towers even when it is not being used to make a call. A phone would stop doing this only if it were turned off or in aircraft mode.

This is the analogy used by the engineers at Inmarsat to explain what happened on MH-370 at the same time that so many other inexplicable events were occurring. The airplane's signal to the satellite stopped, and returned only when the airplane logged back on at 2:25, as if powering on at the beginning of a flight.

There are only a few ways this can be explained: there was a power failure on the airplane, the software failed, or something interfered with the connection between the antenna and the satellite, such as the plane flying upside down so that the fuselage was between the antenna and the satellite. All three possibilities are extremely remote. Some clues, however, have not been pursued.

One week into 2008, a Qantas Boeing 747 was on approach to Bangkok from London with 365 people on board. It was a clear and sunny afternoon—which was fortunate because as Qantas Flight 2 passed through ten thousand feet, it lost electrical power. The autothrottle, autopilot, weather radar, and many other systems, including the automatic control for the pressurization system, simply stopped. Only the captain's flight display worked, albeit in "degraded mode." The plane landed safely, but once it was on the ground, its doors could not be opened because the outflow valves failed to automatically release the cabin pressure.

On the Boeing 747 and other Boeing jetliners, including the 777 and the 767, there is a galley located above the electronics and equipment room, called the E&E bay. On Qantas Flight 2, a flood from the galley above caused water to flow into this area. This was not a one-time event. During its investigation, the Australian Transport Safety Bureau discovered that electronics equipment in the bay had been "repeatedly subjected" to liquid beyond what it was designed to handle. When the ATSB set out to find similar events, it turned up five on large jetliners, four on Boeings, and one on an Airbus A300—and those were just the ones serious enough to have caused a safety event in flight.

I've learned that on airplanes with galleys located above electrical equipment, mechanics often see leaking.

"The 777 has an avionics bay below the first-class galley. When a crew reports water, it is required by the manual to inspect the avionics bay for leaks from water penetration," I was told by a mechanic for a major American airline—I'll call this mechanic Fred because he does not want me to use his real name. A few days after Fred told me this, he sent a video in which I could clearly see water dripping onto the floor of a cramped and noisy equipment room.

"Where did you get this?" I asked, thinking Fred had found the footage on a YouTube-like service for aircraft mechanics. But no, he'd shot the video himself, on a Boeing 767 that came under the care of his wrench shortly after our initial conversation about Qantas Flight 2.

I started to think that maybe some water-induced intermittent electrical problem could have produced the various failures on Malaysia 370, including the puzzling power down and subsequent restoration of communication at 2:25 a.m. that no one has yet been able to explain.

So I asked the ATSB, when Flight 2 lost most of its power on January 7, 2008, did it cause a termination in the link to the satellite? Could this issue with water damaging the electronics affect satellite communication? The ATSB did not know.

"I am unable to provide you with a definitive answer as we would need to establish a detailed understanding of the load-sharing arrangements on the aircraft, interaction with AC BUS 4, not to mention the electrical system that supports the SATCOM," Julian Walsh, the acting chief commissioner, replied in an e-mail. "This was not an aspect of the original investigation," he told me.

I do not know if it is part of the Malaysian investigation into what happened to the plane, which had the registration ID 9M-MRO, because the team doing that investigation does not answer questions.

Once Malaysia 370's last radar echo faded—the one showing it somewhere at the northern tip of Sumatra—Fariq made a final turn. No data suggest when, but the plane turned south and flew on for five hours more until it ran out of fuel. This final turn is the point where I believe Fariq's deprived brain reached its limit. Like Number 14 fixed on the four of spades, Fariq was locked onto some thought. I asked airline captain Pete Frey to try to explain Fariq's last action. I wasn't intending to insult Pete, a longtime friend, by suggesting he might know what it is like to be without cogent thought, and thankfully, Pete didn't take it that way.

"Who knows what he's doing? *He doesn't know what he's doing*," Frey said after considering my question. "He's lost sense of time, so now he thinks he's back there. Maybe he's thinking, 'I've got to head north, and where am I? I'll go this way.' By the time he realizes he's lost, he says, 'Now I'll turn around and go back, but I don't know where I'm going back to, so I'll just head south. I'm too far north.'"

When you consider how muddled Fariq's mind must have been, you can see many ways in which MH-370's bizarre flight path can be explained.

"All you really have to say is at this point," Frey told me, "he's struggling with intermittent abilities, and it's not enough."

Center of Confusion

Commercial aviation is both more and less advanced than people think. The pilot's preflight programming can enable the machine to take off, fly, and land, though any pilot will tell you there's a lot more to the job than just getting the airplane into the sky and back down again (as you will see in part 5 of this book).

"If we have it set up for where the plane is going to fly the flight plan, we could go to first class and have a meal, and it would do those things," said James Blaszczak, now retired, who flew the Boeing 777 for eight years for United before going on to fly the 787 Dreamliner. Yes, that's pretty impressive. Yet at any hour around the world, hundreds of planes are flying isolated in the sky, communicating only sporadically with the ground. In this respect, MH-370 is more like the *Hawaii Clipper.*

The 1930s-era, four-engine Martin 130 flying boat was robust and comfortable enough that up to thirty-two passengers, seated in bamboo and rattan chairs in the lounge, could be served hot meals prepared in the galley by uniformed and gloved stewards. After dining, they could retire to berths made up with blankets and pillows.

A nine-man flight-deck crew was responsible for operating the flight that would get travelers from California across the Pacific to Manila, a five-day journey with overnight stays in island hotels built and operated by the airline.

The crew consisted of a captain, four copilots, a flight engineer, a radio operator, a navigator, and a pilot studying basic navigation from him. Navigation was by celestial observation combined with a calculation of elapsed speed and time, the dead reckoning Pete Frey explained earlier. The navigator could get an assist by using a direction-finding system, antennas that locked onto broadcast radio stations.

"We would tune in certain stations we knew we could use. In those days, one of our favorite things were the high-powered commercial stations along the California coast," said Ed Dover, a radio operator with Pan Am from 1942 until 1948, who later spent thirty-three years as an air traffic controller. "KGO broadcast such a strong signal we could hear them out to sea."

Using an antenna shaped like a figure eight, radio operators like Dover could note the areas of strongest and weakest signal and use that information to determine the direction of the station. "You could draw a line on the map. We knew where the station was; on land, the transmitter was on the map, so we could match the direction in terms of compass direction and say, 'Okay, down that way, that's where we are in relation to where the station was.'" You have to love the term used when navigators and radio operators shared their information to ratify their calculation of position: they called it the center of confusion.

If that kind of direction finding seems like something from the Stone Age, communication technology wasn't much better. The weekly Pan Am transpacific flight from San Francisco to Manila left California on a Wednesday in coordination with two ships leaving on Monday, one from San Francisco, the other from Honolulu. The ships would be midway through their cruises when the Pan Am Clipper passed overhead, giving the airliner a navigational fix and, as significantly, a degree of comfort that someone—even someone eight thousand feet below, in a dark ocean—knew where the airliner was.

Still, darned if those big airships didn't land right where they should have, without a single computer assist.

It is easy to be dismissive of the early rudimentary systems when air travel began. But that a twenty-six-ton flying boat could take off from San Francisco and arrive in Honolulu eighteen hours later was nothing short of a miracle. It was state of the art, as they say.

So people were shocked to learn, four generations later, that with all their fancy navigation and communications equipment, many airlines were still in the dark when it came to knowing exactly where their airliners were when crossing the oceans. As with the Pan Am Clippers, the challenge for overwater flights remains that airplanes are outside the range of land-based radar. If they are going to communicate their position, it will have to be via satellite, which is expensive. As Daniel Baker, the founder and CEO of FlightAware.com explained, the "center of confusion" has given way to the "cone of ambiguity."

An airline sending a position report via satellite every fifteen minutes can cover one hundred fifty miles before sending the next position report, Baker said. Should the location be more precise?

"Satellites don't know where the airplane is. The airplane sends the signal, and that requires the airplane be pointed in the right direction, that is, belly down. If there is an upset on board—a loss of control, the plane is headed straight down or upside down—it can't get a satellite signal, because it is pointed in the wrong direction," he said, adding, "That's a big challenge." And just in case you are thinking, as I did, that he was talking about really unlikely possibilities, he ticked off a few disasters to illustrate his point: EgyptAir Flight 990 and Air France 447. "We are at the limits of technology," he said.

Acausal Connections

Seventy-six years separate the *Hawaii Clipper* and Malaysia Flight 370, and yet we see striking similarities. Both airliners were modern and spacious, and the pilots in command were highly trained and experienced. Upon closer look, however, we see that maybe all wasn't as it seemed. Was Leo Terletsky, forty-three, "one of the most distinguished flight commanders,"* as Pan Am claimed—or was he afraid of flying and so volatile that most pilots were unwilling to work with him, as Horace Brock wrote?

Pilots who flew with Zaharie Ahmad Shah, captain of Malaysia 370, said he was a passionate aviator and a mentor to younger pilots. "A gem of a guy, a real professional, enjoying the best time of his life," one of them told me. Yet a few journalists—quoting unidentified sources, mind you—painted a picture of a political fanatic.

Pan Am's Martin 130 flying boat was a marvel of aviation;

* *Airways*, Pan Am employee magazine, no. 5 (July–August 1938).

custom-built to help the airline span the oceans, but the company's chief pilot in the thirties complained that it was "unstable on every axis and a pig to fly."* While the Boeing 777 airliner is widely used and considered by pilots to be pleasurable and reliable, the list of things that could have contributed to the 2014 disaster exposes unappreciated hazards on the airplane.

If the Pan Am Clipper experienced something catastrophic, the area where it could have gone down is relatively small. If it was hijacked, though, the search zone becomes enormous, because the amount of fuel on board allowed the plane another eleven hours of flight. With Malaysia 370, the plane's satellite indicated the Boeing 777 flew for five hours and fifty-four minutes after mysteriously powering up at 2:25, so predictions about where it ultimately came down could be no more precise than an area from sixty thousand to six hundred thousand square miles.

On a stormy night above the Atlantic Ocean on June 1, 2009, an Air France Airbus A330 with 228 people on board disappeared on a flight from Rio de Janeiro to Paris. Though the plane was five hundred miles outside radar coverage, investigators had a last-known position based on an ACARS message transmitted via satellite. The plane must have entered the water within forty miles of that. Even so, it took five days for the first bodies and floating debris to be discovered, and two years to locate the plane.

Some of the credit for finally finding the submerged airliner goes to Metron Scientific Solutions, a company staffed with pencil-wielding mathematicians who used probability, logic, and numbers to conclude that the likely resting place of the plane was a narrow slice of ocean that had already been checked.

---

* *Flying the Oceans*, by Horace Brock.

"A lack of success tells you about where it is not, and that contributes to knowledge," said Larry Stone, chief scientist at Metron. Talk about having a positive point of view. The Metron method is based on Bayesian probability, the theory of eighteenth-century statistician and philosopher Thomas Bayes, whose first published work, *Divine Benevolence*, was equally optimistic because it attempted to prove that God wants us to be happy.

Using Bayesian logic to look for missing airplanes, as interpreted by Metron, involves taking all kinds of input about the missing thing (even conflicting input) and assigning levels of certainty or uncertainty to each. Everything gets a weight, and everything gets revised as things change. New information, as Stone so cheerily described it to me, is often negative.

In the search for Air France 447, scientists, mathematicians, and underwater technologists were involved in a very difficult bit of detective work. They covered a surface area of eighteen thousand square miles and a debris field nearly three miles deep. Two search seasons after they started, the wreckage of the Airbus was found on the edge of a plain not too far from the beginning of a steep and rugged underwater mountain range.

Many smart people contributed to the search but Olivier Ferrante, then the investigator in charge of the Air France 447 probe for France's Bureau d'Enquêtes et d'Analyses, said they benefited from an additional, highly uncertain element: luck. "The fact that the airplane was on flat terrain" was important to seeing the debris on the sonar pictures, Ferrante told me. "We saw man-made debris, and we identified it in the picture. That was luck. If it had been a couple of kilometers to the east or north, or close to the cliff, we wouldn't have seen it."

I bring up the Air France search in this context because it is the event most like Malaysia 370 in terms of how the newest technological developments are being pushed to do even more within the cone of ambiguity, defined earlier by FlightAware's Daniel Baker as the number of miles a plane can travel between satellite position reports.

While the ACARS messages helped to narrow the area where the Air France 447 airliner might be located, the difficulty in getting a more precise location prompted the satellite company Inmarsat to beef up some of its network by adding new data to the communication transmission. Two additional tidbits of information enable calculations of a plane's location based on how long it takes a message transmitted from the plane to arrive at its destination.

"Inmarsat did modify its systems to add the so-called timing and frequency information to the handshake messages," Ruy Pinto, an engineer and the chief operating officer at Inmarsat, told me at the company's futuristic high-rise headquarters in London. This newly added information would become useful the weekend Malaysia 370 went missing. First, it showed that the plane had not crashed right away, but had flown for hours. Later, the timing and frequency data allowed the company to determine that the plane flew south into the Indian Ocean.

"If MH-370 had occurred at the time of the Air France disaster it would not have been possible to make the analysis that we ended up doing," Pinto told me.

What's missing with Malaysia 370 is even the basic information used by the French, because in the case of MH-370, no ACARS messages were sent after 1:07 a.m. This meant the search area would be massive.

It would be nearly a year and a half before the first debris from Malaysia 370 was found on a beach on Réunion Island, off the coast of East Africa. By then, the wing section had traveled too far and arrived too late to provide any clue as to where the airplane landed in the water. At its smallest, the search area is three times larger than the one in which Air France 447 was found.

Still, the discovery of the wing flap was useful in one way: it shut down the theory that once it disappeared from radar, the plane had turned north, toward Asia and the Caucasus Mountains. One of the more popular proponents of that line of thinking was Jeff Wise, a CNN talking head who wrote a book, *The Plane That Wasn't There*, describing an elaborate plot that required dismissing some of the data from Inmarsat.

"All the inexplicable coincidences and mismatched data went away," Wise wrote about his alternative scenario, getting a big spread in *New York* magazine. "The answer became wonderfully simple."

He was not the only person who thought the plane was hiding in a remote part of the world. Thomas McInerney, a retired lieutenant general and military analyst for Fox News, told the network's morning news program in 2015 that the plane could be in the "the Stans," referring to the countries whose names include the suffix *-stan*. "That airplane can fly nonstop from the Stans to the United States, New York, or Washington, DC. It could be a future trigger for events against the country."

I'll leave that kind of worrying for Fox News watchers. I'm more concerned with a disturbing discovery made while working on this book: for all the apparent effort to try to solve the mystery of MH-370, authorities may not be as committed to finding out what went wrong. That also would not be unprecedented.

Cover-Up

Air crashes have the potential to reveal government secrets and failures, company malfeasance, or all the above. This is even more pronounced if the airline is owned by the government, as is the case in Malaysia and many other countries.

In the case of Malaysia 370, the airline seems to have had a most embarrassing secret to keep: that before the plane flew off into oblivion, the company already knew it was unable to track its airplanes as frequently as required. After the plane went missing at the end of March 2014, twenty-six countries were donating personnel, aircraft, and ships to look for the jet. What would they say if they knew that a year earlier, Malaysia Airline executives had been warned about just this kind of problem? In fact, they had.

In April 2013, and again in June, several company auditors looking at the flight operations discovered a number of problems with the airline's compliance with government and international aviation standards. Most significant was that on

Malaysia's wide-body, long-haul Boeing 747s* and 777s, which would include the plane flying as MH-370, flights could be tracked only every half hour, even though the airline was required to know on a more frequent basis the whereabouts of each plane.

In a presentation to the executives in August, the auditors from Quality Assurance and Regulatory Affairs said that flight following and flight watching could "not be achieved . . . at intervals stated" in the flight dispatch manual. According to the auditors, by law the planes should not have been dispatched.

In a warning that was prescient, the auditors reminded executives that airlines were required to actively watch and track their airplanes "throughout all phases of the flight to ensure that the flight is following its prescribed route, without unplanned deviation, diversion or delay," in order to satisfy government regulations.

So while the rest of the world was shocked that a plane carrying 239 people could just vanish, it could not have been totally surprising to those within the airline's flight operations office, who had been warned seven months earlier.

In the chaotic months that followed MH-370's disappearing act, information about these audits was presented to Hishammuddin Hussein, who was the acting transport minister at the time, by airline employees who were concerned about the implications for future flights. When I was given the audit documents, which are marked "Confidential," I was told it was because Hussein and others in government, even after being told about the issue, had failed to respond.

Malaysia Airlines and the Department of Civil Aviation also

* Malaysia Airlines no longer flies the Boeing 747.

failed to reply to my inquiries about the audits, despite repeated requests.

Certainly, an investigation means the arrival of countless busybodies just like me asking all sorts of questions about what would not necessarily be public information absent an accident.

Another example is the five hundred pounds of lithium-ion batteries that were packed onto the Malaysia flight. There is much interest in whether this highly flammable cargo might have contributed to the crash. The pallets of Motorola walkie-talkie batteries were not declared hazardous because the Malaysians said they had complied with international guidelines for the safe transport of dangerous goods. But had they? And what about the curiously large amount of mangosteens also in transit to China? Some five tons of this tropical fruit were loaded into the hold. No matter how scrumptious they are or how large the appetite for them in China, they are a notably large item on the cargo manifest, because mangosteens were at the tail end of the small and secondary fruiting season in Malaysia, which runs from November to February.

If the plane hadn't gone missing, the contents and veracity of the airline's cargo manifest, and even the airline's failure to meet standards for knowing where its planes are during flight, would have remained the airline's business. Now the world is demanding answers.

How much attention could the airline and the government have been paying if they failed to notice a Boeing 777 flying off course over the country's most populated cities? For all anybody knew, the plane might have been on its way to fly, 9/11-like, into the pride of Kuala Lumpur, the Petronas Towers. Then there is the embarrassing revelation that after MH-370 stopped communicating, it took five hours for controllers to raise an alarm and begin a search for the plane.

Global attention paid to these questions did not bring out the best in Malaysia's leaders, who were alternately confused and combative, and mostly nonresponsive to questions about their investigation. At the time, all information was provided by Hishammuddin Hussein, who had the most reason not to be forthcoming. When the jetliner disappeared, Hussein was not only the acting minister of transportation but also minister of defense. The performance of both departments on March 8 could be kindly described as deficient or, less kindly (if more accurately), derelict. When it came to hyperbole, however, the minister was a master. At one point he described the search as "the most difficult in human history." So one has to wonder if the daily press conferences, conducted in three languages but delivering little new information, were intended to obfuscate, or whether that was just an unintended side benefit.

A lack of transparency leads to one thing: "You're going to have conjecture," explained Jesse Walker, an editor at *Reason* magazine and the author of the 2014 book *The United States of Paranoia: A Conspiracy Theory*. "When you have a blank slate to play with, people will fill in the blanks with stories that interest and excite and feel plausible to them—even apart from the evidence, because this is the way they expect the world to work," he said. "They draw on the narratives they know and find appealing."

After studying a number of accidents with intransigent conspiracy theories attached, I've come to the conclusion that there can be no better tactic for an investigator with something to hide than to encourage those theories. The story doesn't have to be convincing; it need only gain a toehold, after which the public does the rest.

When McInerney, the Fox News commentator, worried about the "Stans" on the one-year anniversary of the loss of MH-370, the ATSB had long ago discounted the Reuters report that the plane had engaged in a series of altitude climbs and descents in an attempt to hide from radar. The length of the plane's flight indicated that it had operated at maximum fuel efficiency. That simply doesn't allow for those kinds of gas-guzzling up-and-down maneuvers, or it would have run out of fuel earlier. Yet there was McInerney, with the credibility of a military general, describing the pilot's intentional acts to the network's 1.6 million viewers.

"He makes his turn at the checkpoint, and all of a sudden he climbs to forty-five thousand feet, which means he's depressurizing the airplane, then he goes down to twenty-three thousand feet and then back to thirty-five [thousand] again," McInerney said, while a simulation of the flight he was describing was shown on the split screen. "He eliminated the people in the back to be a threat," he said, by which I assume he meant to tell his audience that the pilot planned to kill the passengers by depressurizing the airplane. "And now," McInerney continued, "all of a sudden the airplane disappears."

Jesse Walker describes fill-in-the-blanks conjecture as the inevitable result of an information vacuum. The airplane's roller-coaster flight is one example. I'm not suggesting that creating stories was a premeditated effort by Malaysia, but the effect, intended or not, was to create uncertainty about the facts that lingers to this day.

The first major accident I wrote about was the crash of TWA Flight 800 in 1996, which I covered as a correspondent for CNN. My book *Deadly Departure* dealt largely with what caused the plane to explode while climbing out of New York airspace on a flight to Paris. One could not write about TWA 800 without getting into alternative theories, including that it had been shot down by the U.S. Navy or by an Iranian submarine.

On the seventeenth anniversary of the crash, an online television network released a documentary that would, the advertising promo claimed, "blow the lid off an alleged multi-agency cover-up of what really happened." According to the ninety-minute program, objects fired at the plane caused it to blow up, though what, how, and why were not explained. The point of the film was to suggest that two hundred thirty people were murdered and the National Transportation Safety Board and the Federal Bureau of Investigation hid the truth from the public.

The documentary was part of a salvo that included a petition that the NTSB reconsider its probable cause report on the disaster. The safety board had spent four years and twenty million dollars investigating. It had consulted thousands of people from academic, commercial, and research organizations and concluded that the 747's design had allowed the vapors in the plane's center fuel tank to get so hot that the tank could explode. This hazard existed on many airliners during normal flights. From the tests performed by the safety board at a lab at the California Institute of Technology, the NTSB determined that the amount of time a jetliner fuel tank could be in this volatile state, just a spark away from going *kaboom*, was about one-third its operational time.

While the NTSB never determined exactly what set off the blast on TWA 800, the realization that airliners were at considerable risk prompted the Department of Transportation to order changes.

The NTSB denied the request to reopen the TWA Flight

800 investigation, but all the news hype had served its purpose, driving traffic to the online channel where the documentary* was posted.

Deadly Departure was published years before the documentary was produced, and suggested a different kind of cover-up, the cover-up of a design flaw you will read about later in this book and one that is now recognized in the industry. But if you ask people what they remember about TWA 800, most will say something about a missile shooting it down.

---

* Spencer Rumsey, "TWA Flight 800 Exposé Takes Off at Stony Brook Film Festival," *Long Island Press*, July 8, 2013, http://www.longislandpress.com/2013/07/08/twa-flight-800-expose-takes-off-at-stony-brook-film-festival/.

Lost at Sea

Theories about what happened to the *Hawaii Clipper* and Malaysia 370 have another element in common: provincialism. Malaysia's largely Muslim population caused some nervousness in the West that the loss of MH-370 might be the work of Islamic extremists. In 1938, when the *Hawaii Clipper* disappeared, the Japanese were suspected of having played a role.

Between World Wars I and II, the Pacific Ocean was a zone of geopolitical intrigue. America was eager to strengthen its presence in the Pacific, but was prevented by the 1922 Washington Naval Treaty from any military buildup on the islands west of Hawaii. So when Pan American Airways applied to the U.S. government to develop facilities for civilian seaplane service on Midway, Wake, Guam, and the Philippines, the airline's and the military's interests were aligned. Pan Am would build bases complete with radio stations, power plants, fuel supplies, maintenance operations, and housing, so it could have stopover points for its transpacific flights. After the treaty restrictions expired in 1936, the military could take advantage of that infrastructure.

America was also eager for land-based airfields in the region, a project of aviation pioneer Gene Vidal, director of America's Bureau of Air Commerce. Vidal had already participated in the creation of three airlines—Eastern, Trans World, and Northeast—and he thought that flying boats had a limited future. In 1935 he oversaw the colonization of three small islands that could provide southern Pacific air access to Australia, New Zealand, and Singapore, and he arranged federal funding for an airfield to be built on one of them, Howland Island.

The first use of this airfield would be to provide a refueling stop for the upcoming around-the-world journey of Amelia Earhart. The much-anticipated flight provided perfect justification for building a runway with fuel and service support in an area where Japan had a strong presence.

Combine the star power of Angelina Jolie with the ceaseless trending of the Kardashians to get an idea of the level of celebrity adoration Earhart enjoyed during the 1930s. She was a woman in what was perhaps the first extreme sport of the industrial age.

Sure, within the small world of early aviators, there was grumbling about her flying skills, her occasional errors in judgment, and her pushing on when waiting might have been wiser. She also never became fluent with the radio, which was critical because it was more than a form of communication; it was a source for determining direction. Still, these are ordinary shortcomings. Amelia Earhart made history because of her extraordinary strengths, tenacity and fearlessness among them. And she married a man willing to do more than support her unconventional career. George Putnam was Mr. Amelia Earhart, devoted to promoting the First Lady of Flight.

Earhart was the first woman to fly across the Atlantic Ocean and the first pilot to cross the Pacific solo from Honolulu to the U.S. mainland. While she was not the first woman pilot, she was the one who most famously used her career in the sky to promote women's equality on the ground.

Always pushing to accomplish one more feat, Earhart set out from Miami on June 1, 1937, to circle the globe at the equator. It was an undertaking punctuated with problems, one of which was that her navigator, Fred Noonan, had a drinking problem. Well into the endeavor and one week before crossing the Pacific, the most difficult part, she complained about Noonan's drinking to her husband. In his book, *Amelia Earhart: The Final Story*, author Vincent Loomis said Earhart told Putnam on June 26 that Noonan was "hitting the bottle again, and I don't even know where he's getting it!"

Noonan had been Pan Am's chief Pacific navigator, a licensed pilot and a master mariner, charting and mapping the routes pilots would use to Asia. He navigated the *Hawaii Clipper* on its first flight with fare-paying passengers. So his drinking must have been pronounced for it to have cost him his job at Pan Am, which several accounts claim it did. Yet his leaving Pan Am made him available for Earhart's round-the-world flight.

Noonan's direction-finding skill was top-notch. It would need to be, because the last portion of the exhausting trip would be the 7,000-mile transpacific flight to Vidal's new airstrip on Howland Island, one by two miles in size. From Lae, New Guinea, it was a 2,556-mile, eighteen-hour journey.

After two days preparing the plane and waiting for suitable weather, Earhart and Noonan took off from the jungle runway at Lae at 10:00 a.m. on July 2. In her diary of the trip, published posthumously as *Last Flight*, Earhart wrote that she was still hoping to arrive in California in time for the Fourth of July.

Earhart and Noonan flew through the day and into the night, crossing the international dateline and passing a ship that confirmed to radio operators on the ground that the plane was right on course.

At 2:45 a.m., the radio operator on the U.S. Coast Guard cutter *Itasca*, which was waiting by the island to guide Earhart's Lockheed Electra to land, began receiving radio messages. By 7:42 a.m., which was close to Earhart's estimated arrival time, the messages were getting disturbing. *We must be on you but cannot see you. But gas is running low. Been unable to reach you by radio. We are flying at altitude one thousand feet.* The next was also troubling: *We are circling but cannot hear you.*

Earhart and Noonan never found Howland Island. Their last message was sent at 8:44 a.m.

By circumnavigating the equator east to west, Earhart knew she had left the most difficult part of the trip for last. "Howland is such a small spot that every aid to locating it must be available," she wrote. She would be "glad when we have the hazards of its navigation behind us."

Earhart was not expressing just her own concerns. During the planning stages, others advised that her plan was not workable. Navigator, explorer, and Harvard professor Brad Washburn spent an evening reviewing details with Earhart. He worried about her trying to find Howland without radio signals to home in on. Mark Walker, the Pan Am first officer who would go missing on the *Hawaii Clipper* one year later, said the challenge was insurmountable. In a letter to *Shipmate*, the alumni magazine of the U.S. Naval Academy, Walker's cousin Robert Greenwood wrote that early in 1937, Walker was assigned by Pan Am to help Earhart and Noonan in planning the Pacific phase of the trip. Greenwood said his cousin urged Earhart not to risk "such a foolhardy publicity stunt," and that "her equipment was barely adequate."

In addition to his concern that Earhart was ill-equipped, Walker was especially troubled about Japanese kidnappers and hijackers. Okay, maybe he was a worrywart. He told his younger sister, Mary Ann Walker, that his role in protecting newsreel footage taken of a Japanese air attack on the U.S. gunboat *Panay* in 1937 had resulted in personal threats from the Japanese. Even so, he was not the only person to imagine a sinister turn of events when Earhart failed to arrive at Howland Island.

Earhart and Noonan had personal histories that, from a skeptic's point of view, suggest that they could have been carrying out a secret agenda during the flight. Charles Hill, author of *Fix on the Rising Sun*, proposes that Earhart defected to Japan, delivering Noonan, his valuable navigational skills, and insider knowledge of Pan Am's Pacific air routes as a gift to the Japanese. Other theories suggest the opposite: that Earhart was an American spy, having accepted a mission to fly over the Japanese islands in the Pacific to photograph them and assess troop buildup.

Regardless of how preposterous the theories, they demonstrate that people were already on edge and prepared to believe practically anything when the Pan Am *Hawaii Clipper* vanished in the Pacific just as Earhart had less than a year earlier. Beyond the basic facts of the flight (the who, the when, and the where), there was little information about what happened to the flying boat, and that fueled speculation.

Horace Brock was in Manila awaiting the arrival of the *Hawaii Clipper*. As he tells the story in his autobiography,

Flying the Oceans, when he heard that the plane was overdue and probably down, he took a cab to the U.S. Army air base at Clark Field. He barged into the commander's office, wanting to know why B-16s weren't being sent to search for the missing plane. The commander was apologetic but firm, according to Brock. "Son, my men have families, too, wives and children. They have no navigation experience. I doubt if any one of them could find his way back." Flying over vast amounts of ocean in the decades before GPS was not for the fainthearted.

As far as Brock was concerned, a combination of bad weather and Captain Terletsky's lack of the right stuff was to blame for the tragedy. All over Manila, however, Pan Am crews were accosted by people insisting that the Japanese had taken the plane.

The search for the Clipper was still under way on August 4, 1938, when the Hearst International News Service reported a blockbuster story. FBI agents had been undercover at the Pan Am Alameda base since January in an attempt to "thwart any sabotage" at the company and "protect the nation's most ambitious private air route." A memo to FBI director J. Edgar Hoover dated February 5, and obtained through a Freedom of Information Act request by Clipper researcher and documentary filmmaker Guy Noffsinger in his ongoing effort to find out what happened, confirms that seven months before the *Hawaii Clipper* went missing, the bureau was investigating the possibility of vandalism of Pan American Airways flying boats. After the loss of the Martin 130, the acting secretary of commerce sent a letter to Hoover thanking him for information "relative to possible sabotage in connection with Pan American Airlines [*sic*] ships." Through it all, Pan Am was circumspect; "all lines of merit" were being investigated, the company told the Hearst reporter.

The elaborate theories about the *Hawaii Clipper* didn't get fleshed out for many years, and when they did, it was almost by accident. In 1964 an Earhart searcher named Joseph Gervais was investigating whether the wreckage of an old airplane on the Pacific island of Truk was Earhart's Electra. It was not, but having come five thousand miles, he had nothing to lose by sitting down with the locals and listening to their stories. According to them, fifteen people arrived on the airplane before the start of the war, and they were escorted by the Japanese, who were using Truk as an air base. The travelers were executed, and their bodies buried below a concrete slab.

Remarkably, after hearing the story, Gervais's response was "I'm not interested in a plane with fifteen people. I'm interested in a plane with two people, a man and a woman." Then, in 1980, he had a change of heart after reading Ronald W. Jackson's *China Clipper*, a book that sets the loss of the *Hawaii Clipper* against the backdrop of the conflict in the Pacific.

The Japanese were apprehensive about the development of Pan Am's island seaplane bases. They understood how the American military could use them to dodge international prohibitions on arming in the Pacific. When the first Pan Am survey flight from San Francisco to Honolulu was flown, a Japanese newspaper editorial noted the worry: "Even if the route is restricted to commercial flights, who can assure that it would not be used for military purposes in case of emergency?"

Jackson writes that the Japanese set out to disrupt the Clipper service. On the eve of November 22, 1935, the inaugural transpacific flight, FBI agents arrested two Japanese nationals who had slipped on board the *Hawaii Clipper* as it sat in the harbor at the Alameda base, across the bay from San Francisco. The men had been tampering with the plane's radio direction finder, key to navigating across the vast ocean. The airline kept the arrest quiet. On January 5, 1936, as another Pan Am captain was sailing the same flying boat through a channel in San Francisco Bay, the hull was sliced by several concrete pillars studded with iron rods that sat just below the water's surface. Who had placed them there was not known. Once again, the suspected vandalism was hushed up.

From these accounts, Jackson concluded that the *Hawaii Clipper* was hijacked by Japanese stowaways who'd boarded the plane on its overnight in Guam. Backing up his scenario is an FBI report from William L. MacNeill, a former U.S. Marine who worked for the military and Pan Am in the village of Sumay for three years in the mid-1930s. MacNeill claimed that a spy ring operated in Guam and that the Japanese had "all the chance in the world to plant a time bomb on any ship or clipper that comes in." All this convinced Gervais that the Pan Am mystery was worth another look. His narrow focus on Earhart broadened.

I pause here to point out that ten years before deciding he had stumbled on the final destination of the *Hawaii Clipper*, Gervais cowrote* *Amelia Earhart Lives*, a book claiming that she survived capture by the Japanese during the war and afterward returned to the States to assume a new identity as Irene Bolam. Gervais met Bolam, a private pilot in her youth, through a mutual friend at a flying club meeting in Long Island. Bolam was said to look a lot like Earhart, leading Gervais to believe that she actually *was* Amelia Earhart. There were several problems with his theory, the most significant of which was that Bolam insisted she was not Earhart. So bear this in mind when I refer to Gervais's findings.

---

* With Joe Klaas.

In 1980, Gervais was invited to meet with a group of retired World War II–era Pan Am mechanics who would view the photos of the airplane he saw in Truk sixteen years before and consider the possibility that the passengers and crew of the *Hawaii Clipper* ended up interred there. The sixteen former Pan Am men quickly concluded that the plane was not a Martin 130, and two days later they sent a recommendation to airline management advising against an investigatory visit to the island, according to documents in the airline's historical archives at the Richter Library at the University of Miami.

To me, the meeting with Gervais gives the impression that the company was trying to get at the truth, but in a memo to executives, Pan Am's then-director of corporate public relations, James Arey, wrote that the official position had not changed. As ever, "the Clipper was lost during a storm."

Ten years earlier, however, Pan Am founder and chairman Juan Trippe had a very different view.

In a memo Guy Noffsinger found in the airline's archives, Harvey L. Katz, who preceded Arey in the Pan Am public relations department, details a meeting with Trippe on August 26, 1970, at which the recently retired former CEO dropped this startling tidbit. "Mr. Trippe said that after the war, he was told by the Navy department that the Japanese hid in the aircraft and commandeered it in mid-flight," wrote Katz. There was more: "The aircraft then was flown to a Japanese base where the engines were studied and, according to Mr. Trippe, were copied in detail for use on Japanese fighter aircraft. He said passengers and crew were killed." Katz wrote the memo to John C. Leslie, a senior VP for Pan Am international affairs.

So why was the airline not receptive to Gervais's account? The crash hunter's mistake, according to Charles Hill, was hanging the entire hijacking theory on the aircraft wreckage he photographed in Truk in the 1960s. The Pan Am review committee said it was a British Sunderland, a four-engine flying boat of the same era used by the Royal Air Force during the war. The misidentification of the plane allowed the committee to dismiss all Gervais's claims.

Hill's *Fix on the Rising Sun* is at times complimentary to Jackson's and Gervais's accounts and at other times contradictory; and the book is often incomprehensible. It does include some of the same details Trippe revealed, and Hill's theory is eerily similar to the one proposed by Jeff Wise in the disappearance of Malaysia 370.

Both these armchair investigators, Hill and Wise, proposed that the flights were skyjacked by technically savvy interlopers who took control of the planes and then made deceptive transmissions. With the *Hawaii Clipper*, the theory goes, hijackers forced the pilots to fly to a Japanese-controlled island. In Malaysia 370, the hijackers were Ukrainian, and the destination was Russian-controlled Kazakhstan. In Hill's version of events, the crew was desperate to communicate their plight surreptitiously to ground stations while under the watchful eye of the Japanese skyjackers—so Pan Am's radio operator, McCarty, transmitted false navigational fixes, a kind of code that, when deciphered, would point the listener to the location of three Japanese seaplane bases in the Pacific. The message being "The Japanese have us."

In the Malaysia 370 scenario Wise proposed (before the discovery of the wing flap convinced him he was wrong), the skyjackers slipped through the floor hatch leading to the aircraft's electronics bay near the cockpit. Having accessed the plane's satellite data unit, they reprogrammed it to transmit signals that would send out phony information about where the plane was heading, sending the search-and-recovery teams on a wild goose chase on the wrong side of the equator. With access to the plane's brain, the hijackers seized the flight controls from the pilots and remotely flew the plane to their target destination.

In the case of the *Hawaii Clipper*, was McCarty's ingenious code lost on the recipients? Did the airline's ground operators figure out the message but were subsequently told to keep its significance to themselves? Hill does not say. The official statement was then, and still is to this day, that no one knows what happened to the plane—just as with Malaysia 370.

"I'm considered among the whack-a-doodles," Wise told me when journalists were still interested in what he called his "spoof theory" of MH-370. A charmingly self-effacing science writer with a private pilot's license and a penchant for the technical, Wise is not a basement-dwelling nerd spinning plots involving hostile foreign powers and Ernst Blofeld–style, computer-hacking villains. Well, maybe he is a little, but Wise didn't expect the world to embrace his view. He just wanted the possibility considered.

I never bought Wise's "north to Kazakhstan" idea, but we did agree on one thing. The most troublesome piece of data, the one that opens the door to consideration of some kind of hacking into the electronic system, is that after MH-370 disappeared from radar, the signal to the satellite was inexplicably lost for as long as an hour and twenty minutes. Something interfered with the satellite data unit, or SDU. This is why I was so interested in the water damage to the E&E bay on the Qantas flight to Bangkok in 2008 and in similar events.

"Nobody has tried to grapple with the key data point, the

reboot of the SDU," Wise told me, noting that deactivating the satellite communication system is not something most pilots would know how to do. To Wise, this cries out for further study. "This indicates to me that there was tampering by somebody, and a tampered piece of equipment—you have to put an asterisk by that."

Wise's spoof theory required imagining a state-backed plot involving a number of people with detailed knowledge of the inner workings of a sophisticated, computer-driven machine. It points out the alarming possibility that airliners can be digitally commandeered. In this respect, Wise is the voice of a small community of people who warn that it is indeed possible. The digital airliner has outpaced the industry's ability to protect against all cyber threats.

In a presentation to the 2014 Black Hat, a computer security conference, Madrid-based cyber security expert Ruben Santamarta demonstrated how he hacked into an airliner's SDU through Inmarsat's SwiftBroadband connection. Santamarta said he was able to bypass normal security gates and log on using the industry standard naming protocol of the aircraft.

Once in the satellite data network, a hacker can "modify settings, reboot the terminal, turn off the terminal, and do nasty things," he told the crowd. "Obviously we are not crashing airplanes with these vulnerabilities, that has to be clear," Santamarta said. "These attacks—one can be used to disrupt or to modify satellite data links."

Santamarta's claim has been challenged by those who argue that his lack of access to actual equipment used by airlines casts doubt on his conclusions. Still, if it could theoretically happen, it must be considered a hazard; one the industry should be dealing with—yesterday.

Wise is the lucky alternative theorist whose scenario has been disproven to his satisfaction. Being wrong isn't too bad if it provides closure. It is a different story for Guy Noffsinger. "I'm in it for the long haul," he told me well into the second half of a decade spent wandering down the many side roads of the Pan Am Clipper enigma. He is waiting for some kind of satisfying ending, which, like the missing flying boat, is nowhere to be found.

PART TWO

Conspiracy

Just because you're paranoid doesn't mean they aren't after you.

—JOSEPH HELLER, *Catch 22*

A Little Mistrust

If Juan Trippe, with all his political connections, knew that the *Hawaii Clipper* had been taken by the Japanese in 1938, choosing to share it in a conversation with a public relations executive in 1970 was a very understated way of setting the record straight. So while his astonishing confirmation of the long-held theory about the *Hawaii Clipper* added another curious element to the story, it fell short of providing certainty.

Ah, certainty. Before I started writing this book, I had no idea how elusive certainty could be in investigating air crashes. Yet the more accidents I looked at, the more odd elements I found.

The Helios 522 case seemed straightforward until 2011, when consultants hired by a Helios mechanic and three executives of the airline, all facing criminal charges in the accident, reexamined the wreckage as part of their defense and came to a conclusion that differed from the official report. With the assistance of Ron Schleede and Caj Frostell, now retired from the International Civil Aviation Organization, they asked to reopen the investigation.

Schleede and Frostell had questioned whether the pilots failed to pressurize the aircraft, based in part on their examination of the system-selector knob found at the scene in the Off position. When the consultants examined it, they thought the scoring on the back was evidence it might have been pushed to the Off position upon the airplane's impact with the ground, rather than because the pilots failed to pressurize the plane on takeoff. That would have meant a case of mechanical failure rather than pilot error. Boeing disagreed; the Greeks and the Americans opted not to reopen the case. The official report had been published; public attention had moved on.

An air disaster dominates the news until the next story. For those involved, however, the investigation, in all its gritty, tedious detail, is enormously important. People can face jail, as the Helios workers did. Airlines and manufacturers can be sued and fined, ordered to make expensive design or operational changes, and subjected to new regulations. Aviation authorities can be exposed as derelict, and government secrets can be exposed.

In air accident investigations in some countries, unlike in criminal cases, people with an interest in the outcome take part. The airline, pilots, maintenance workers, air traffic controllers, flight attendants, product manufacturers, and government officials work together. The idea is that that their conflicting interests keep them all in check.

Still, there is a real knowledge disparity. For example, when Inmarsat arrived in Malaysia with the news that its satellite data could be used to help locate the airplane, its calculations showed that the 777 flew into the South Indian Ocean. This news was met with skepticism.

Inmarsat's vice president for aviation, David Coiley, kept defending the company's research, telling me that the calculations and conclusions were peer reviewed. But seriously, who were the company's peers? This was new to everybody. Coiley said even his own people didn't understand completely. "We could tell [only] so much from a simple handshake or logon."

An experienced tin kicker is a generalist—a mile wide and half an inch deep. Conversely, the designer of a microprocessor or satellite communication system is a specialist—half an inch wide and a mile deep. The trend for future investigations will be toward more sophisticated, niche areas of specialization, according to Robert MacIntosh, former NTSB chief of international aviation affairs. "We're going to have to depend more and more on the technical expertise we get from the manufacturers."

Cue the menacing *Jaws* theme music here, because this approach calls for trusting the untrustworthy, those with a stake in the outcome, said Florida State University professor Lance deHaven-Smith. A fervent contrarian, deHaven-Smith said people who will suspect the activities of foreign governments are reluctant to doubt their own, even though they should. "We got enough events where the government is not giving us an adequate explanation," he said. When accidents happen that benefit the powerful, or happen with a frequency that defies the odds, a little mistrust can be a good thing.

There were many reasons to question the cause of the crash that killed Dorothy Hunt, wife of White House fixer E. Howard Hunt, in 1972. Ms. Hunt was a former CIA operative who was said to have delivered hush money to the Watergate burglars in the scandal that led to President Richard Nixon's resignation.

On an overcast afternoon in December of that year, she was flying on United Flight 553 from Washington to Chicago.

Snow and freezing rain fell as the Boeing 737 stalled on approach to the city's Midway Airport and plowed into a residential neighborhood. The NTSB said in its final report that it found no evidence of "any medical condition that would have incapacitated the crew, or of any interference with the crew in the performance of their duties," in short, no evidence of foul play. Still, there were several eyebrow-raising details. Hunt was carrying $10,000 and had bought $250,000 of flight insurance before boarding.

"It's pretty wild when you have the White House being blackmailed by a former CIA agent to keep quiet about a crime," deHaven-Smith said of the circumstances. "Then she dies in a plane crash with ten thousand K? If you're not suspicious of that, you're crazy."

The investigation was more difficult because the flight data recorder was not working. Despite that, the NTSB found oversights by the crew during the critical period as the plane neared the airport.

By analyzing the engine noise and other sounds captured by the cockpit voice recorder and time-syncing them with the air traffic control radar, the investigators deduced that the pilots were trying to comply with requests from ATC to slow their arrival so that another plane could clear the runway. With gear down and spoilers deployed, the crew did not maintain enough speed after leveling the airplane and got dangerously close to a stall.

Flight 553 was at one thousand feet, just below the minimum decision height, when the controller asked the crew to execute a missed approach. At precisely that time, the stick shaker started and the pilots retracted the flaps to fifteen degrees and applied takeoff power.

The 737 descended through the cloud cover in a level attitude and then quickly went nose high as it slammed into a number of homes, killing two residents inside one bungalow and forty-three people on board the airplane.

Charles Colson, once special counsel to Nixon and another character who was jailed for his role in Watergate, would later tell *Time* magazine that Dorothy Hunt was murdered by the CIA. The charge remains alive among those of a conspiratorial bent. Yet as a murder plot, it falls short on the credibility scale.

Too many people would have had to be involved to carry off a complex plan that also had to factor in the unforeseeable conditions that would lead to the plane's speed getting away from the pilots. Murder by airplane is a concept more likely to succeed in crime fiction than in reality.

In the movies, the bad guy tampers with the victim's car, which then goes off a cliff. In aviation, however, sabotage must do more than create the failure mechanism; it must make sure it goes undetected while triggering the catastrophe at just the right time so that it can also penetrate a highly developed safety net. Short of commandeering the airplane and purposely crashing it, detonating a bomb, launching a missile, or setting a fire—all of which leave evidence in the wreckage—intentionally causing a crash is not so simple.

"What's possible and what's not?" asks retired airline pilot and novelist John Nance, whose books sometimes feature crime at thirty-five thousand feet. He knows how tricky it is to come up with a credible murder plot where an airliner is the weapon. "The linchpin is predictability; how certain are you that A is going to produce B? That's what you must have."

Parts can be tampered with to create a crime novel plot, such as slicing the brake line, but in aviation, the most capricious elements of all, according to Nance, are sitting at the front of the airplane. Pilots can rise to the challenge and save the day or they can founder and become another link in an unbroken chain to disaster. There are fascinating examples of both pilot heroics and failures, which you will read about later in this book. Yet rare is the would-be assassin who can orchestrate all the instruments of destruction in advance. Sometimes an air accident is just an accident. Other times, however, it is an enigma.

A Diplomat Dies

One thing investigators don't expect to find on the scene of an air disaster is passengers with gunshot wounds. Yet two of the fifteen people on the plane with United Nations secretary-general Dag Hammarskjöld had been shot, and that was just one of many surprising discoveries. There are various theories about what caused the plane to fly into the trees on a dark night in September 1961. It could have been an assassination or kidnapping plot or an attempted interception by mercenaries to divert Hammarskjöld from his peace mission. It could have been a mechanical problem or an error by the crew. Although the accident has been investigated four times, what really happened remains a mystery.

The UN-chartered DC-6 was approaching the airport at Ndola, in Northern Rhodesia,* during a violent interlude in the decolonialization of the Congo. The Belgians had ostensibly pulled out of the country, but in the resource-rich state of

---

* Now Zambia.

Katanga, European-backed mercenaries were still around supporting its attempt to secede from the Republic of the Congo.* Hammarskjöld wanted a cease-fire between the mercenaries and the UN troops that were there to assist the Republic of the Congo. Stopping the violence was to be the first of a two-step effort to reverse Katanga's secession. Because of their commercial interests in the region's resources, Africa's colonial powers, Belgium, Britain, and France, opposed the UN plan. The Americans had a different concern: in these Cold War days, certain factions in the U.S. government worried that the Soviet Union would take advantage of the turmoil in Africa to gain an advantage.

So when Hammarskjöld died in a plane crash, it was like an Agatha Christie novel. There were plenty of suspects.

Accompanying the secretary-general on the flight were two UN executives, four security officers, two soldiers, and a secretary. Six men, all Swedes, made up the flight crew.

For security reasons, Hammarskjöld's trip to Northern Rhodesia to meet with Katanga leader Moise Tshombe was hush-hush: The plane would follow a circuitous route. The crew would maintain radio silence, using only an emergency channel staffed by an operator communicating with them in their native Swedish.

But if the Hammarskjöld visit was a secret, it was badly kept. Journalists, protesters, and mercenary pilots were waiting at the airport, along with Lord Cuthbert Alport, the region's British high commissioner. Tshombe was there, too, under a special exception to the whites-only rule imposed in British-controlled Northern Rhodesia.

---

* Now the Democratic Republic of the Congo.

The Swedish DC-6, known as *Albertina*, was on final approach to the Ndola Airport at around midnight on September 18. During the last bank, a wing hit the trees and then the ground not far from a twelve-foot anthill. The plane plowed into it and cartwheeled to the right. The still-spinning propellers on the right wing dispersed fifteen hundred gallons of fuel along a three-hundred-foot path as the plane slowed and came to rest. There was a fire on the ground, but it could not be determined when it started. Evidence, autopsies, and eyewitnesses offered conflicting information. The plane could have been on fire as it flew; it could have caught fire when it crashed; or the fire could have been rekindled after the crash. There was testimony to back up all three possibilities.

The captain of the *Albertina*, Per Erik Hallonquist, had radioed the tower of his anticipated arrival at 12:20 a.m. Why the air traffic controller waited until 2:20 to issue an alert when the plane did not arrive is not clear. Maybe he was reassured by Lord Alport, who for some reason told the tower staff not to worry about the missing plane because Hammarskjöld had probably changed his plans.

That night, John Ngongo and Safeli Soft were camping in the forest not far from the airport, tending a charcoal-making kiln. It was clear and sometime after 10:00 p.m., Ngongo said, when he and Soft saw a large aircraft fly overhead, followed by a smaller plane that sounded like a jet. The engine and the wings of the big plane were on fire, according to Ngongo. The two men got to the crash scene at dawn, where they found the plane smoldering and the body of a man, whom they later learned was Dag Hammarskjöld, lying apart from the plane, propped up against an anthill.

Susan Williams, author of *Who Killed Hammarskjöld?*

The UN, the Cold War, and White Supremacy in Africa, writes that the men went to Timothy Kankasa, the township secretary, to report what they had found. Kankasa went to the scene with them and returned to call the police. Kankasa said it was between 9:00 and 9:30 a.m., but it wasn't until afternoon that he heard ambulances. By contrast, a number of other locals say they came upon the wreckage that morning and that it was surrounded by uniformed soldiers and police and cordoned off with red tape.

Obviously, it couldn't be both. The Federation of Rhodesia authorities' story doesn't match either of the bystanders' accounts. The authorities said the plane wasn't discovered until 3:15 p.m., fifteen hours after it crashed eight miles from the airport. The mysteries were multiplying.

If Kankasa called the police in the morning, why did it take them so long to get to the site? If police were there early on, what were they doing and why was 3:15 p.m. given as the time of their arrival? These are not academic questions because there was a surprise in the middle of the charred wreckage: a survivor.

Harry Julien, Hammarskjöld's director of security, was suffering from first- and second-degree burns, sunburn, a fractured ankle, and a head injury, but he was very much alive. If the authorities had deliberately slowed getting him medical treatment, why?

"We know that the crash was known; Timothy Kankasa reported it early on to the authorities, but nobody came," Susan Williams told me. "The authorities knew about the crash. We know there were people there. We know the ambulances didn't come." The sun was blazing, it was September, and Julien had been suffering for fifteen hours.

"I am Sergeant Harry Julien, security officer to the OUN," he told his nurse at Ndola Hospital. "Please inform Léopoldville of the crash. Tell my wife and kids I'm alive before the casualty list is published."

Harry Julien entered the hospital with a good prognosis, but six days later he succumbed to renal failure. A. V. (Paddy) Allen, the police inspector who accompanied Julien to the hospital, was surprised because he did not think Julien's injuries were life-threatening, adding another odd element to an exceptionally odd case.

In the ambulance, Julien shared details of the accident with Allen and later with hospital personnel. A tape recorder in his room was supposed to document what he said, but either the machine was not turned on or the tapes disappeared, because the reports contain only the brief statements Julien made to the police, the nurses, and physician Mark Lowenthal.

"It blew up," he said, when asked what happened as the plane was making its pass over the runway. "There was great speed. Great speed."

Officer Allen asked him, "What happened then?"

"Then there was the crash," Julien replied.

Julien said the others were trapped on the plane. Autopsies showed they were all badly burned. This made it even more curious that Hammarskjöld's body, untouched by fire, was outside the airplane when witnesses saw it. The diplomat could have been tossed out of the plane and away from the blaze on impact, or someone might have moved him.

Hammarskjöld was not the only one found in an unexpected condition. One passenger was in the cockpit, an unusual place for him to have been on approach to landing, and then there were the two who had bullet wounds.

The Federation of Rhodesia and Nyasaland Commission of Inquiry under the director of civil aviation concluded that the bullet wounds had to have been the result of fire detonating the ammunition being carried on the plane. The experts consulted by Williams for her book concluded that this was not possible. "Ammunition for rifles, heavy machine-guns and pistols cannot, when heated by fire, eject bullets with sufficient force for the bullets to get into a human body," according to a Swedish explosives expert. Another said, "If bullets were found in the bodies of any of the victims of the air crash, they must have passed through the barrel of a weapon."

As to the cause of the crash itself, a dozen scenarios have been considered. In his book *Disasters in the Air*, Jan Bartelski weighs the theories and suggests one of his own.

A nearly undamaged instrument panel from the captain's position was found in the wreckage, and the static line to the altimeter was disconnected. In their report, the Rhodesian investigators discounted the significance of this, but Bartelski suggested that the bad instrument could have led the pilots to believe they were higher than they were as they approached the airport.

As in my MH-370 theory, Bartelski acknowledges certain assumptions. His scenario is based on his experience flying the DC-6. According to Bartelski, *Albertina* was a DC-6B, which had an idiosyncratic pressurization system. It did not always depressurize on landing. On one occasion a crew member was blown out the door because of the positive pressure differential—in a situation like what happened to American Airlines purser José Chiu in Miami in 2000. For this reason, pilots flying the DC-6B depressurized the aircraft before landing, at an estimated two thousand feet.

Capt. Per Erik Hallonquist arrived at the Ndola airport ten

minutes earlier than he anticipated and had to make a steeper and faster descent than planned, dumping the cabin pressure at a higher altitude by opening the emergency pressure-release valve. This sudden, drastic change "could have had a catastrophic effect on the flexible line" to the altimeter, Bartelski writes. Getting from a separated static line to making the approach sixteen hundred feet too low involves a number of missteps, combined with design features unique to the DC-6 and certain laws of physics.

"I'm not saying it couldn't have happened that way; it could," said Nick Tramontano, who piloted and worked on DC-6s during his career with Seaboard World Airlines.* When I asked him to analyze the altimeter-gone-bad theory, he said he'd take a look and compare it with DC-6 maintenance manuals.

Tramontano said that the pilot's and first officer's separate altimeters could have given the same and, in this case, incorrect altitude information if the captain switched the static source to alternate, the position in which the switch was found at the crash site.

Bertelski writes that investigators were not as familiar with the specifics of the DC-6B as they should have been, which caused them to dismiss this important evidence in the wreckage.

With the exception of Bertelski's detailed explanation, nearly every other theory has a malevolent element, in keeping with the violent and chaotic last days of colonialism in central Africa. Mercenary pilots have confessed to, or boasted in public about, shooting down the plane. These comments were plausible. On the flight just before the one to Ndola, *Albertina* had been hit with machine-gun fire by Katangan forces in

---

* Later Flying Tigers.

Elizabethville,* and the damage repaired. People at the crash site saw holes in the fuselage consistent with weapons fire. The Rhodesian inquiry, however, concluded that shots to the plane would not have disabled the flight controls enough to cause it to crash.

A separate team assembled by the United Nations was working at the same time as the Rhodesians. The UN team hired Max Frei-Shulzer, a Swiss microbiologist and forensic scientist, to answer the question of whether the plane had been attacked.

Why Frei-Shulzer was hired is puzzling. His work with the Zurich Police Department involved handwriting analysis and lifting evidence from surfaces with tape—think fingerprints and fiber residue—but he had no expertise in aviation or metallurgy. His technique for settling the issue of whether bullets were in the fuselage consisted of melting down four thousand pounds of fused aluminum and looking for the presence of other metals. Bartelski describes it as "an extremely critical metallurgical process requiring accurate temperature control." After Frei-Shulzer made soup of the plane, he said he could "exclude the possibility of hostile actions from the air and from the ground." He also discounted sabotage.

In 1962, the Rhodesian board of inquiry followed Frei-Shulzer's lead, concluding that the crash was "probably due to human failure." The complete discounting of overt action against the flight was not accepted by the UN then, or in its 2015 review. Yet these and other contradictions have been examined multiple times as people associated with the case have come forward with new information and as forensic tech-

* Now Lubumbashi.

nology advances. Still, some unexplored aspects of the accident cannot be reviewed half a century later.

For example, it would not be possible to reconstruct some of the critical factors impacting the pilots' performance, MacIntosh, the retired safety investigator, said. "The issue of making this nighttime approach, where everybody is looking out the window and nobody is looking at the altimeter, and you think you are flying level and you are not, those things are never going to be discussed." After flying multiple missions in the Congo for the U.S. Air Force, MacIntosh left the area two weeks before the Hammarskjöld crash. His interest in what happened has not waned.

Yet the person responsible for the most recent revival of the Hammarskjöld whodunit is not a scientist, aviation professional, or criminologist. Susan Williams is a British historian specializing in Africa. On every research trip to the continent, she would come across some thread of the mystery that begged to be pulled. And she pulled enough of these threads to weave her own complex tapestry that includes independent analysis of memos, witness statements, and photographs of the bodies. She makes a compelling case that during the conflict-ridden period marking the beginning of the end for colonialism in central Africa, what happened to the secretary-general was probably deliberate. The cover-up was possible only because the Federation of Rhodesia and Nyasaland was ruled by Britain, which was able to control every aspect of the investigation.

National, racial, political, and commercial biases all played a role, she said. To try to extract those intrusions from what is supposed to be the objective work of air crash investigators would be like trying to reassemble the *Albertina* after it was boiled down by Frei-Shulzer.

The Hammarskjöld Commission relied heavily on Susan Williams's work *Who Killed Hammarskjöld?* when it prompted the United Nations to launch the fourth inquiry into the crash early in 2015. After a review of the evidence that lasted nearly a year, a three-member panel of experts made one small step forward when it ruled out an uncontrolled descent and, therefore, a midair explosion. Controlled flight into terrain was the most likely scenario, the panel said, based on the swath of downed trees along the final flight path and the wreckage distribution. This small advance does not explain what caused the pilots to fly the plane too low. Still, Williams finds it heartening.

"The truth is starting to emerge, and I find it exciting," Williams told me, adding optimistically, "It would be hard to imagine that a cover-up like this could happen now."

Susan, read on.

The Dodge

On either end of 1985, air safety agencies in Bolivia and Canada were thrust into one of America's biggest political scandals when plane crashes in those countries were linked to wide-ranging Reagan administration programs to provide support to Nicaraguan rebels and arms to Iran in the Iran-Contra Affair.

American lives were lost on American-made airplanes operated by American airlines, but America's air safety agencies did not participate in the investigations in a significant way, and a probable cause was not satisfactorily determined. Two books suggest cover-ups intended to hide the relationship of the airlines to secret and perhaps prohibited U.S. government activity.

On December 12, 1985, a DC-8 operated by the Miami-based air charter company Arrow Air crashed fifteen seconds after takeoff from Gander, Newfoundland. Two hundred fifty-six people were killed, most of them U.S. Army soldiers from the 101st Airborne Division, the Screaming Eagles. The soldiers were on their way home for Christmas to the base at Fort Campbell, Kentucky.

In December 1985, Arrow Air wasn't just moving soldiers; it also had contracts to transport high-explosive incendiary ammo and forty-millimeter shells and weapons, according to Richard Gadd, who coordinated the movement of weapons for the CIA.

"There were layers and layers of intrigue associated with the crash," said Les Filotas, one of ten members of the Canadian Aviation Safety Board. There was evidence of mechanical, operational, and possible criminal elements in the accident. What Filotas found bizarre was that staff investigators at his agency didn't seem too interested in pursuing these bewildering leads.

"There was a lot of transportation of weapons from Egypt to the U.S. or from the U.S. to Egypt," Filotas told me, an allegation he made in *Improbable Cause*, his book about the bureaucracy and politicization of the Arrow Air crash investigation. Filotas couldn't say if the plane's flight history or the airline's relationship to the CIA were relevant to what happened, but he thought both should be checked out.

Under international agreements, it was the responsibility of the Canadians to look into the circumstances, but on either side of the border it was obvious this was a sensitive case. The flight was a U.S. military charter on an airline associated with the CIA. The cargo included mysterious boxes whose contents were never identified. In the cargo hold and the passenger cabin—if past flights were any indication—there was a supersize collection of weapons and ammunition, as soldiers stashed these kinds of combat souvenirs in their bags. Furthermore, the plane had been left unattended during extended stopovers in Egypt and in Cologne, its intermediate stop on the way back to the United States.

All this was going on during a sensitive time in American and Iranian relations, as the United States was negotiating for the release of hostages held in Lebanon. National security aide Oliver North had ticked off Iran by selling it weapons that were not the ones the Iranians expected. In a memo to Robert McFarlane and Richard Poindexter, North's bosses at the National Security Agency, he urged them to fix the problem. Failing to get the Iranians the missiles they wanted would "ignite Iran fire—the hostages would be our minimum losses." Was the Arrow Air disaster a fulfillment of that warning?

What prevented a detour into the briar patch of spies and soured secret deals was the quick dismissal of terrorism, sabotage, or even an in-flight explosion on Arrow Air Flight 1285. After the disaster, claims of responsibility came from the group Islamic Jihad and the Independent Organization for the Liberation of Egypt, but the Canadian air accident investigator and a spokeswoman for the government were dismissive right away.

"A lot of groups will claim responsibility, and every claim will be looked into," Helene Lafortune, from the Canadian Department of External Affairs, told *The New York Times*. "They use that to promote their cause. I don't think it's a lead on anything," she said of the groups claiming responsibility.

Compare that to the October 2015 in-flight breakup of a Russian charter flight leaving the resort town of Sharm El Sheikh in Egypt with a load of holidaymakers headed back to St. Petersburg. The Metrojet Airbus A321 came apart and crashed into the Sinai Desert, killing 224. While American, British, and Russian politicians were quick to say the plane was felled by a bomb, Egyptian investigators insisted that there was no evidence of this. The Islamic fundamentalist group ISIS claimed to have been responsible for the disaster. Despite Egypt's reticence, news organizations reported as fact that the plane was bombed out of the sky.

In Canada thirty years earlier, however, all deference was paid to the official investigation. A team of thirty-two soldiers from the U.S. Army, under the command of Maj. Gen. John Crosby, arrived in Gander, as did a group of FBI agents. While the American soldiers were permitted to retrieve the victims, the FBI agents were confined to their hotel rooms by the Canadians. The Royal Canadian Mounted Police did not need the FBI's help to search for evidence of a crime.

The Canadians were handling the air accident investigation without America's help also, though the NTSB's George Seidlein was on hand as a designated representative. Soon, however, he would be replaced by the agency's chief of major investigations, Ron Schleede, who said he flew to Gander with someone from Pratt and Whitney at the request of NTSB chairman Jim Burnett. "Our chairman never trusted anybody," Schleede told me. "He wanted a second set of eyes, and I was sent to sort that out."

One early theory was that ice on the wings may have prevented the plane from climbing, sending it slamming into the trees less than a mile past the airport. Filotas wasn't yet a member of the Canadian Aviation Safety Board (CASB), but he had some idea of how this theory might have been generated, even though no ice was seen on the wings by the six people who serviced the plane before its departure, and only a small quantity of snow grains was observed at weather observation sites at the airport.

"I can imagine how this huge accident happens in Gander, and our board chooses investigators who might have investigated small accidents. They go down there. They meet the engineers of Pratt & Whitney, and they all have a meeting," he told me, describing the organizational session that kicks off any investigation. The airline, the airplane, engine makers, unions for the crew, and government aviation organizations are all represented.

"They're swamped by experts with a point of view, and they come up and say, 'It might have been an engine failure,' and the engine maker says, 'No, it's impossible.' And it goes around the room. 'Could have been ice?' And they all nod their heads [and] say, 'Yeah, it could have been ice.' Our investigators can be overwhelmed, and they convince each other."

Schleede told me that when the crash happened, an FBI agent hurried out to his car to drive to the airport and discovered that his windshield wipers were frozen to the glass. To Schleede, this was notable; the ice theory wasn't just a construct. Schleede told me, "I'm willing to admit when I'm wrong," but when it came to supporting the Canadian ice theory, he insists he was right.

It was an opinion that had him in conflict with Seidlein, who didn't think ice caused the crash. Seidlein, who died in 2008, was sent back to his office in Chicago, never to return to Gander, never more to champion the theory that contradicted that of Schleede and the tin kickers at CASB.

Filotas doesn't know how the ice idea started, of course. His round-the-table "Yeah, it's ice" scenario was speculation during one of several lengthy phone conversations about his book. Based on early news reports, ice on the wings at takeoff was mentioned as a possibility within hours of the crash, even before the first organizational meeting. Anyway, Filotas was being charitable about the initiation of the ice theory. On other matters, he was far more conspiratorial.

There were behind-the-scenes discussions between CASB and the Americans that he was not privy to, he tells me. He is politically savvy enough to conclude that "General Crosby was all over the crash site talking to investigators. And they don't have to tell anybody to hide anything. Somebody from the Justice Department just has to phone the chairman and say, 'It would be inconvenient for us if one of your board members starts shooting his mouth off about a bomb.'"

Filotas was so convinced that ice wasn't the cause that he and three fellow board members wrote a dissent to CASB's official probable cause. Filotas was an aeronautical engineer, as were members Norm Bobbitt and Dave Mussallem; member Ross Stevenson was a former DC-8 pilot with Air Canada. (A fourth man, Roger Lacroix, a former combat pilot and a member of the Royal Canadian Mounted Police, disagreed with the ice theory but resigned from the board before the final report and dissent were released.) When these men received the preliminary staff report in 1987, they found an overreliance on the scant data from the flight recorder. The witness statements were missing, and recordkeeping from the engine inspections was shoddy.

This deep level of scrutiny by political appointees who were not air safety specialists cannot be called unprecedented, because CASB was too new to have much of a history. It was created only a year before the Arrow Air accident, in an attempt to separate safety from regulation. Ironically, the controversy over Arrow Air and another crash in 1989 would prompt a second overhaul of the way Canada handled air accidents.

In the second half of the eighties, however, CASB's civil servants were not expecting to be cross-examined by board members over engineering minutiae such as engine teardown procedures and the limitations of data gathered from four channel flight data recorders.

"There was an ongoing dispute at the board level about what the role of the board member was," Peter Boag, the man in charge of the CASB investigation, told me of Filotas and the others who challenged the staff view of the evidence. "They wanted to be investigators, but that wasn't their role."

There's a lot of he said/he said at this point. Filotas claims to have begun a review of the staff's findings only because he found misrepresentations in it. Perhaps the investigation was too large and complex for Boag to handle, Filotas thought. On that impression, he was not alone.

Peter Boag had a "very thin background, extremely thin," the NTSB's MacIntosh said, calling the investigator in charge of the Arrow Air crash overconfident to a fault. "I can't imagine how you could approach him and get constructive suggestions going when his attitude toward everything was 'I can handle it, and I've done a good job.'"

During one particularly testy exchange recounted in *Improbable Cause*, the members promoting a theory other than ice were asking Boag for the plane's maintenance records from an earlier accident when Boag made it clear he had had enough. He stood up to leave, saying, "Frankly, gentlemen, the well is dry for me. I've done all I can do."

"Once the idea [of ice] got permeated, it was convenient, and all the departments thought it was a good idea," Filotas said. He was troubled that the official investigation had disregarded eyewitnesses' accounts and toxicological exams that suggested a fire or explosion.

Four bystanders who saw the DC-8 in its brief flight across the sky said they saw fire or a glowing light on the aircraft before it slammed into the woods. Yet information from these witnesses was dismissed. When CASB chairman Ken Thorneycroft explained why to Lynn Sherr for ABC News's *20/20*, it sounded as if the board were cherry-picking witnesses.

"We have taken evidence from two hundred or three hundred witnesses; obviously we can't call them all," the board chairman told Sherr. "So let's have a roundtable discussion, decide what evidence we want to have come out at the public hearing, and then select a group of witnesses who will provide that evidence."

The most disturbing contradiction concerned the autopsies. The bodies were taken to the Armed Forces Pathology Lab in Dover, Delaware, and examined by Dr. Robert McMeekin. He declared the deaths instantaneous, but no separate determination of cause was given beyond "plane crash." This misnomer was applied to all 256 victims, regardless of individual injuries.

The toxicology results produced by the Canadians showed that more than half the victims had carbon monoxide or hydrogen cyanide in their systems, which indicates smoke inhalation. This meant that "the soldiers were alive, and they breathed carbon monoxide. There was either a fire or something aboard the plane that emitted carbon monoxide prior to the plane crashing," wrote Cyril Wecht in his book *Tales from the Morgue* after examining the medical records. "These soldiers had to breathe in the CO while still onboard the plane because the autopsies show that the soldiers were dead on impact."

This is the "obvious evidence that this airplane was on fire and it came apart" that board member Roger Lacroix was talking about when he disparaged the ice-on-the-wings scenario in his interview with *20/20*.

In the final report, however, Dr. McMeekin's time-of-death determinations changed. They were no longer instantaneous.

Now death was estimated to have occurred between within thirty seconds to five minutes of the crash for 125 of the victims, which explained the presence of fire-related chemicals in 62 of them. The investigators' facts were changing to preserve the theory that there was no fire before the crash, Filotas said.

When he arrived in Gander for a private planning session with the CASB in the spring of 1986, the NTSB's Schleede had been instructed by NTSB chairman Jim Burnett about what to focus on: Arrow Air and FAA oversight of the airline. It may have seemed like a small matter then, but a 1989 House Judiciary Committee subcommittee hearing noted that of all the government agencies that should have been involved, it was left to the tiny and obscure NTSB to be the fig leaf for the U.S. position. "All the government agencies deferred to the NTSB. The NTSB deferred the question of terrorism to the Canadians," the subcommittee concluded.

"I was told by Burnett to be a hard-ass on Arrow and FAA," Schleede told me. Arrow Air was "a shabby operation," and the FAA hadn't been supervising it sufficiently. These issues appeared in CASB's majority report in December 1988. This was the report that the U.S. congressional subcommittee claimed did not receive sufficient scrutiny by the NTSB.

It doesn't take an overly active imagination to conclude that the United States was diverting attention away from the clues that the plane might have been lost because of a fire or explosion resulting from weapons or sabotage or a terrorist act. This paved the way for CASB's majority report finding that the plane had stalled on takeoff, probably from ice on the wing and a loss of thrust on the number four engine. It was the position Boag and his investigators had taken almost from day one.

Like ice crystals on a windshield in a Newfoundland winter, the ice deniers clung to their position in a dissenting report that said that a fire broke out while the plane was in flight.

Seven months later, Willard Estey, a retired Supreme Court justice in Canada, reviewed the reports and found both of them incredible. "Surmise and speculation inside and outside these proceedings abound, factual evidence does not. Nothing indicates any hope of uncovering explanations of this accident in those areas."

The judge's finding didn't dissuade Boag. A lawyer is looking for something different from an accident investigator, he told me. "I wasn't there, and you weren't there, and the people who were are dead," Boag said. "There could have been complicating factors. The potential power loss from a compressor stall in the engine might have been a complicating factor. Ice represented the most likely, the best if not a definitive finding as to cause."

Filotas called Estey's conclusion that a cause would never be found circular logic. "We shouldn't go on with the investigation because we wouldn't be able to find the cause, and dropping the investigation would ensure that we never did," he said.

And that's the way it was left.

This airplane was not lost in the sea or downed in a global hotspot. It crashed in a First World nation less than a mile from the airport, carrying American servicemen and women home to their families.

The United States "had to know what happened," Filotas said when we talked in 2015. He said it was only the second year since 1985 in which no journalist had been in touch to ask him about the case.

"The military had this tragic loss, and here we were bumbling along. If the U.S. didn't know [what happened to the plane] would they have let this controversy 'ice/no ice' alone or would they have tried to get involved?" He insisted, "They don't care what we do here because they knew what happened."

Judge Estey seems to have been correct in his dim prediction for finding the truth about Arrow Air. Filotas was equally convincing when he said that what the congressional subcommittee hearing called "a near total absence" of a U.S. investigation can only be seen as willful. Why wouldn't a country pull out all the stops in trying to uncover every detail of an accident with such a high toll unless the truth threatened some great scandal?

When Eastern Airlines captain George Jehn first delved into the controversy over the Arrow Air disaster, he was stunned by the similarities with an accident earlier that same year. "The Reagan administration had a definite modus operandi for handling potentially embarrassing air disasters," he told me. Jehn is convinced the mysterious crash of a Boeing 727 in Bolivia was another disaster the U.S. government didn't want to solve.

Eastern Airlines Flight 980 crashed on a mountain just outside La Paz, Bolivia, on New Year's Day 1985, killing twenty-nine people. By international agreement, the country where an accident occurs is in charge, which is why Canada handled the Arrow Air case. For Eastern Flight 980, it was a Bolivian investigation. Still, because everything about the accident except the location was American, people from the NTSB, Boeing, Eastern, and the airline's unions had a right to participate, and to some extent they did. A number of them went to Bolivia, but little of what they learned there shed light on what happened.

Barry Trotter, Eastern's then accident investigator, flew down to South America along with the airline's chief pilot and the VP of flight operations.

"We didn't sit around," Trotter told me. "We went to where the aircraft departed from to get information from ATC and the radar people, the en route radar people. I had a rep with me from the NTSB who was specialized in ATC centers and towers and so forth."

Trotter was talking about Michael O'Rourke, the NTSB's air traffic control expert. What O'Rourke found on visiting the tower at La Paz was a post–World War II–era setup. He was very concerned about the age of the equipment and the lack of maintenance.

"I insisted the FAA come down and flight-check all the nav aids," O'Rourke told me, remembering the on-site visit quite clearly. The equipment had been installed nearly forty years earlier and had never been checked. "It blew my mind," O'Rourke recalled.

In a story in *The New York Times* after the accident, reporter Richard Witkin zeroed in on this issue, writing that immediate attention would be paid to the "accuracy of navigation aids" on which the crew would have relied during the plane's descent over the mountains and on its approach to the airport. Witkin was partially correct. The FAA did test the airport with a specially equipped Boeing 727 and found everything in order, but there was no report on whether the equipment was accurate on the night of the accident.

In his book *Final Destination: Disaster*, Jehn writes that between New Year's and the FAA flight test a few weeks later, repairs had been made to the system. "A rep was there" from the company fixing things before the inspection, Jehn told me. "Of course it was going to pass."

The first attempt to get investigators to the site came about five days after the crash, when helicopter pilot Rus Stiles took a Sikorsky Black Hawk over Mount Illimani. The helicopter was on loan from United Technologies. The company's chief executive, Harry Gray, was a personal friend of Ambassador Arthur Davis, whose wife was killed in the disaster.

"All we could see was a cut in the snow that formed the shape of the wing," Stiles told me. "The plane was so buried, I don't remember any part of the airplane. I didn't see the tail at all." Stiles said he could hover, but landing or leaving people on the mountain was out of the question. Nearly a year would pass before investigators attempted to access the wreckage again.

Jehn read all the early news accounts and was struck by how many leads there were to follow: criminal, political, and operational. Initially, there was reason to suspect an in-flight explosion because people who lived nearby reported hearing "a roar of thunder" and seeing pieces of the plane falling from the sky. Another news story raised the possibility that Davis, the U.S. ambassador to Paraguay, might have been the target of an assassination attempt by the country's military dictator, Gen. Alfredo Stroessner.

Ambassador Davis was scheduled to be on the flight with his wife, Marian, but changed plans at the last minute. Davis and Stroessner had a sometimes contentious relationship, and there was more: Paraguay was earning the ire of the U.S. government for its involvement in drug trafficking. Two months before the crash, the Americans slashed financial aid to Paraguay, and the bulk of the remaining three-million-dollar appropriation went to the Peace Corps. By coincidence (or maybe not), the director of the Peace Corps in Paraguay was also killed on Flight 980.

Eastern Airlines' participation in the movement of illegal

drugs also had to be considered relevant, Jehn said. The airline had been in the news for repeated violations of drug trafficking laws. Drugs had been seized twenty-two times on Eastern's jets in 1984, and the contraband was always hidden in areas accessible only by aviation personnel. One pilot testified to a Senate committee during a hearing that he saw money being taken off planes at Eastern's Panama hub.

Based on the stories pilots told him and also reported to the FBI in Miami, Jehn suggests that Eastern Airlines may have been assisting the Reagan administration's clandestine effort to supply weapons to Nicaraguan rebels in the Iran-Contra, arms-for-hostages, drugs-for-rebels, it-has-a-lot-of-names scandal that dominated the fortieth president's second term from 1984 to 1988.

It wasn't all political intrigue, though. Like the questionable integrity of the navigational aids at La Paz, Jehn and the ALPA (Air Line Pilots Association) pilots who were part of the probe discovered alarming facts related to human error. None of the three men in the cockpit that night had any experience flying in South America, with its high terrain and low-tech navigation facilities. Company policy required a check captain to supervise the first flight into the region, but not the second. This trip back to Miami was the pilot's second flight; the supervising pilot was riding as a passenger.

Don McClure, an ALPA air safety pilot, flew a re-creation of Flight 980 and noted that the airline's radio-based navigational system, called Omega, "continually steered the aircraft off course approximately five miles to the east," or toward the direction of Mount Illimani. Jehn found errors on navigational charts.

Jehn's interest in Eastern Flight 980 was professional: an

airliner had crashed; colleagues were dead. Working separately on a parallel course was Judith Kelly, whose mission was more personal. William Kelly, her husband of sixteen years, was the Peace Corps director in Paraguay. He was traveling aboard Eastern 980 to Miami on business. Judith, who also worked for the Peace Corps, had stayed behind in Asunción. It didn't take her long to realize that her government wasn't pulling out all the stops to determine what had happened.

Six months after she was made a widow, Judith Kelly flew to Bolivia, hired a guide, and climbed Mount Illimani, arriving at nineteen thousand feet to the place where the wreckage of the plane was scattered. She took a few small pieces, left letters she had written to her husband, and climbed back down. Then she traveled to the United States and appeared on NBC's *Today* show to talk about what she'd done and to challenge the NTSB to get up there and have a look. "I made it, a woman, on my own," she explained to Jehn when the two met years later. Her message to the NTSB: certainly the U.S. government and Eastern airlines, with all their resources, could get up there and conduct a proper inquiry. She'd proven it could be done.

From her home in San Antonio, Texas, Alisa Vander Stucken, twenty-eight, watched Judith Kelly berate the NTSB on the morning news in the summer of 1985 and thought "it was awesome." Her husband, Mark Louis Bird, was the second officer on Flight 980. Like Kelly, Vander Stucken was discouraged, and puzzled at the lack of progress in the probe. "I expected both the government and the airline to get down there and investigate and find out what happened," she said. But after the crash faded from the news, Vander Stucken's only source of information was in the letters and phone calls from Judith Kelly. "I think it's pretty sad when a woman has to go up and try and do that for herself," she said of Kelly's hike up Mount Illimani, "instead of the airline or the government; the NTSB. I mean, that's what they're there for."

In his book, Jehn claims that then NTSB chairman Jim Burnett, a Republican appointee of President Reagan who died in 2010, was accommodating the interests of the White House or Eastern or both with this perfunctory investigation. No one I spoke with who was associated with the NTSB then agrees with that conclusion. Peter Kissinger, the board's managing director at the time, said he could not imagine why there would have been reluctance to look into the circumstances. "We truly live by the adage no stone unturned," he told me.

Still, Jehn's perception that Burnett was extremely political did square with that of others who spoke to me both on and off the record. Ron Schleede recalled disagreeing with Burnett over the handling of another accident investigation. "He wanted to have a hearing for some political reason," and Schleede was opposed, he explained. Ultimately Schleede prevailed, but "Burnett was pissed," the investigator said. "He said, 'Okay, can we get Mr. Schleede reassigned to the railroad division?'" It sounded like a joke to me, but that's not how Schleede saw it. "A lot of senior people left the NTSB because of Jim Burnett," he said.

At the time of Flight 980, the NTSB did have a close relationship with Eastern. Burnett's special assistant was a former Eastern pilot named John Wheatley. "Burnett, for some reason, he was really more in contact with Eastern than the other carriers," said Tom Haueter, the former director of aviation safety at the NTSB. "Why that was, I don't know, but it did seem to me that there were a lot of Eastern people around at the time."

Eleven years after the crash, NTSB investigator* Greg Feith

* Feith retired from the NTSB in 2001.

told a meeting of the International Society of Air Safety Investigators that politics can drive a case in one direction or another. He knew about it firsthand because when the NTSB finally acted on Judith Kelly's challenge, Feith was chosen to lead an expedition to the wreckage on Mount Illimani.

In late August 1985, quite out of the blue, Feith was part of an NTSB teleconference in which investigators were invited to volunteer to climb to nineteen thousand feet and look for the black boxes of the Boeing 727. Feith was just twenty-eight and new at the job. He lived in Denver, at an altitude of five thousand feet. His workouts kept him in good shape, and included hiking in the Rockies. Recognizing it would be good for his career, he volunteered to go. More experienced investigators had also stepped up, but it was Feith who was selected. Feith didn't know that the agency's interest in the accident had been dormant, but he knew the NTSB was being pestered by Judith Kelly.

The first indication that something was not right with his assignment was the timing. The climbing season, May to September, was over. It had been ten months since the crash, but all of a sudden it was rush, rush, rush. "It was thrown together at the last minute." Feith told me he was notified on September 25, arrived in DC on October 1 to plan the trip, and left for La Paz on October 2. "There wasn't a lot of planning time. It was more or less, 'You've got the go-ahead. See ya.'"

Feith was joined by two engineers from Boeing, Jim Baker and Al Errington, and two representatives of ALPA, pilot Mark Gerber and his brother Allen, who was not a pilot. The four men were experienced climbers. Royce Fichte, a diplomat at the American embassy in La Paz, and Feith would follow their lead and that of their professional guide, Bernardo Guarachi, who took Kelly up Illimani in June and had been the first to inspect the crash site on foot, in January, at the request of Col. Grover Rojas, the Bolivian air force's director of rescue operations.

The men had little time to acclimate even to the twelve-thousand-foot altitude of La Paz, and the lack of preparation was apparent immediately. The climb began on October 8 and was beset with problems. Errington succumbed to altitude sickness first, so he and Baker did not go on. The Gerber brothers, Fichte, and Feith continued up the mountain, but then Mark Gerber fell ill and Fichte also began to deteriorate. Feith made it to the crash site, but said he had just hours there to dig around in the snow. Though he found the plane's tail section, where the flight and data recorders should have been, he did not see the black boxes. Fearing for the health of the team, the group started back down the mountain to La Paz the following morning.

It was a frustrating experience. Errington suggests that the men were too eager, especially the experienced climbers. "We guided the team through youthful overexuberance," he told me. They were feeling good enough to push forward instead of remaining cautious. "We should have gone slower, and we realized it, but we really wanted to get things done. We could have slowed things down, but we didn't."

The Gerber brothers interpret the time pressure and the assortment of other planning and support blunders as a deliberate campaign to foil them. "The whole situation was rather bizarre," Mark Gerber said. "Looking back, we were there to do a job, yes, but it was a sham." The Gerbers agreed with Jehn. "There were political games going on," Gerber said.

Supporting their suspicion is this: even before the climbers left for La Paz, the Bolivian Board of Inquiry on Accidents and Incidents had written its report. It had been submitted on

September 4, and concluded that "the accident was apparently caused by the aircraft's deviation from its airway." It was a statement of the obvious that didn't include any of the information the NTSB had gathered. Of course it was a deviation from the airway—the airway didn't cut through the mountain. The question was why the plane was where it should not have been. If there was any effort made to find the answer, it wasn't in the report. The issues raised by McClure, the findings of O'Rourke, the reports of eyewitnesses—none of that made it in, either.

Considering how much information has come from armchair investigators, it is fitting that after thirty-one years, two adventurers from Boston should be the ones to claim to have discovered pieces of the black boxes. In May 2016, Dan Futrell and Isaac Stoner climbed Mount Illimani to a field of debris where they found uniforms, engines, human remains, and fragments of orange metal. On their return to the U.S., Stoner posted on the internet site Reddit, "There's a smashed up flight recorder on my kitchen counter!" Michael Poole, once the head of Canada's flight recorder laboratory, said data might be recoverable even after all this time. Even so, the NTSB had no plans to reexamine an accident that was the responsibility of the Bolivians, a spokesman said.

The inconclusive conclusion of the Bolivians is the only official record of a crash that was in every way an American catastrophe, just as Arrow Air had been. Just as with Arrow Air, the NTSB was satisfied to let the Bolivians have the only and final word.

In his role as chief international aviation affairs officer for the NTSB from 1988 to 2011, Robert MacIntosh navigated the complex and often delicate relationships between governments.

Even he was surprised at the noticeable lack of interest in pursuing the safety and security lapses exposed by the parties looking into the crash of Eastern 980. "It's not typical, and that's about all I can say."

Around Christmas 1985, as the first anniversary of the crash approached, a journalist asked the NTSB investigator in charge, John Young, about Flight 980. "Any secrets about the crash are buried beneath snow at an elevation where excavation is virtually impossible," he said.

Young, who died in 2005, might as well have written the script for many investigations to come, including MH-370, where reliance on recovering the airplane is so great that even the Royal Malaysian Police can't make progress without it. When asked for the status of the police probe into the disappearance of MH-370, Inspector-General Tan Sri Khalid Abu Bakar replied, "I'm not at liberty to reveal" any news "until at an official inquiry when the black boxes are discovered."

While everyone agrees that having the airplane is nice, not all investigations are so neatly presented. Without ever leaving his desk, George Jehn came up with enough clues to write a book filled with possibilities. Investigators never know what they're going to find when they start asking questions and digging through records. And before they give up, first they have to try.

PART ONE: MYSTERY

Reporters fill the ballroom used for press conferences at the Sama-Sama Hotel at Kuala Lumpur International Airport. | *(Photo courtesy the author)*

From left: Malaysia Airlines CEO Ahmad Jauhari Yahya, acting minister of transport Hishammuddin Hussein, and director general of civil aviation Azharuddin Abdul Rahman at a news conference in Kuala Lumpur on March 31, 2014. | *(Photo courtesy the author)*

The 9M-MRO, an eleven-year-old Boeing 777, on approach to Los Angeles International Airport three months before it disappeared. | *(Photo courtesy Jay Davis)*

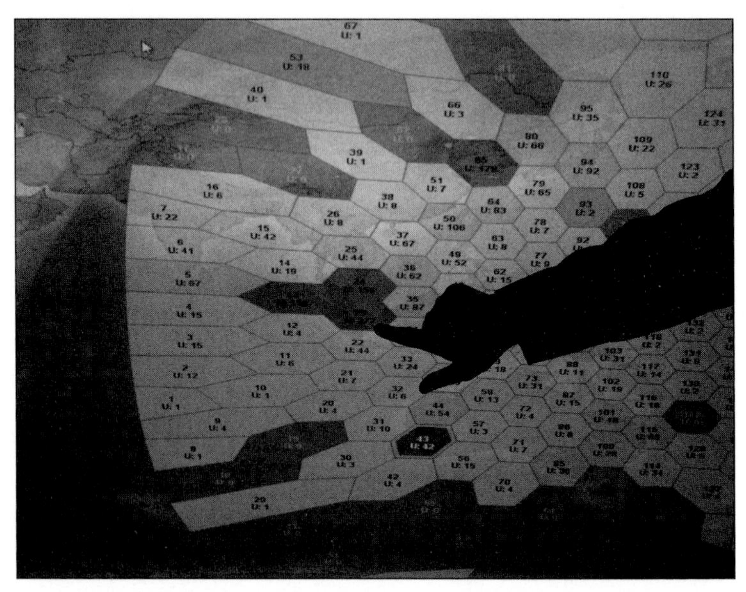

The region of Western Australia where the search for MH-370 was centered. Data from Inmarsat's satellite network led the company to conclude that the plane flew south, which assisted in the search effort. | *(Photo courtesy the author)*

Wreckage of GOL Airlines Flight 1907, which crashed after colliding with a private jet. | *(Brazilian Air Force via Creative Commons)*

PART TWO: CONSPIRACY

The *Hawaii Clipper* on the day it was christened in Pearl Harbor, Honolulu, May 3, 1936. Nine-year-old Patricia Kennedy (*seated on the plane, left*) poured coconut water over the flying boat's bow. Two years later the plane disappeared on a flight from Guam to Manila; it has never been found. | *(Photo courtesy Pan Am Historical Foundation)*

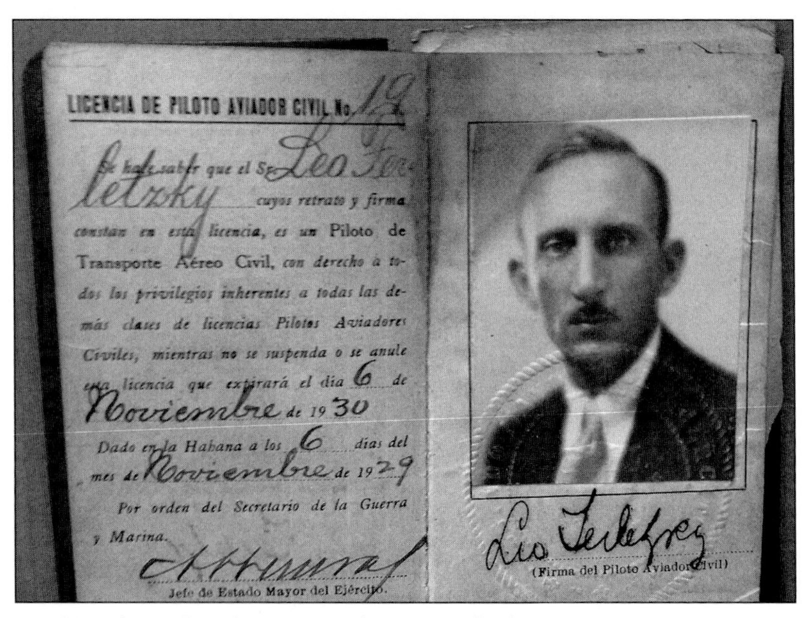

LICENCIA DE PILOTO AVIADOR CIVIL No 19

Se hace saber que el Sr. Leo Terletzky cuyos retrato y firma constan en esta licencia, es un Piloto de Transporte Aéreo Civil, con derecho á todos los privilegios inherentes a todas las demás clases de licencias Pilotos Aviadores Civiles, mientras no se suspenda o se anule esta licencia que expirará el dia 6 de Noviembre de 1930

Dado en la Habana a los 6 dias del mes de Noviembre de 1929

Por orden del Secretario de la Guerra y Marina.

Jefe de Estado Mayor del Ejército.

(Firma del Piloto Aviador Civil)

The Cuban pilot's license issued to Leo Terletsky in 1930. He was captain of the ill-fated *Hawaii Clipper*. | *(Pan Am Historical Foundation / courtesy University of Miami)*

The tail assembly of Helios Flight 522 rests on a hillside. | *(Photo from the Air Accident Investigation and Aviation Safety Board)*

An unidentified official at the scene in the forest where the Douglas DC-6 came down on September 18, 1961, killing the UN secretary-general and fourteen others. The sole survivor of the crash died in the hospital six days later. | *(Photo copyright Adrian Begg and used with permission)*

Snow on Mount Illimani, in Bolivia, shows signs of where Eastern Flight 980 hit the mountain on New Year's Day 1985.
| *(Personal photo courtesy Rus Stiles)*

The crash of an Arrow Air DC-8 in Gander, Newfoundland, on December 12, 1985, remains unsolved, and the official report remains controversial. | *(Royal Canadian Mounted Police photo)*

Two hundred fifty-six people were killed in the Gander crash, most of them members of the U.S. Army's 101st Airborne Division, the Screaming Eagles, who were on their way home from overseas duty. | *(Canpress photo by Jann Van Horne)*

The Air New Zealand DC-10's tail-mounted engine lies not far from the jet's landing gear in the snow on Mount Erebus, in Antarctica. | *(Photo courtesy New Zealand Police)*

PART THREE: FALLIBILITY

Lt. Thomas E. Selfridge (*left*), seated next to Orville Wright, moments before Selfridge died in the crash of the Wright airplane undergoing trials at Fort Myer, Virginia, in September 1908. | *(Photo by Carl H. Claudy, National Air and Space Museum Archives)*

G-ALYP is greeted enthusiastically when it becomes the first jetliner to carry passengers. On January 10, 1954, it would break apart and fall into the sea near the Tuscan island of Elba. | *(Photo courtesy British Aerospace Hatfield)*

At the Royal Aircraft Establishment in Farnborough, England, workers test a Comet fuselage in a water-filled tank. | *(Photo courtesy British Aerospace Hatfield)*

The captain of ANA Flight 692 addresses passengers after the emergency evacuation of the 787. | *(Photo courtesy of passenger Kenichi Kawamura)*

ANA Flight 692 on the tarmac at Takamatsu Airport after its emergency landing. | *(Photo courtesy of passenger Kenichi Kawamura)*

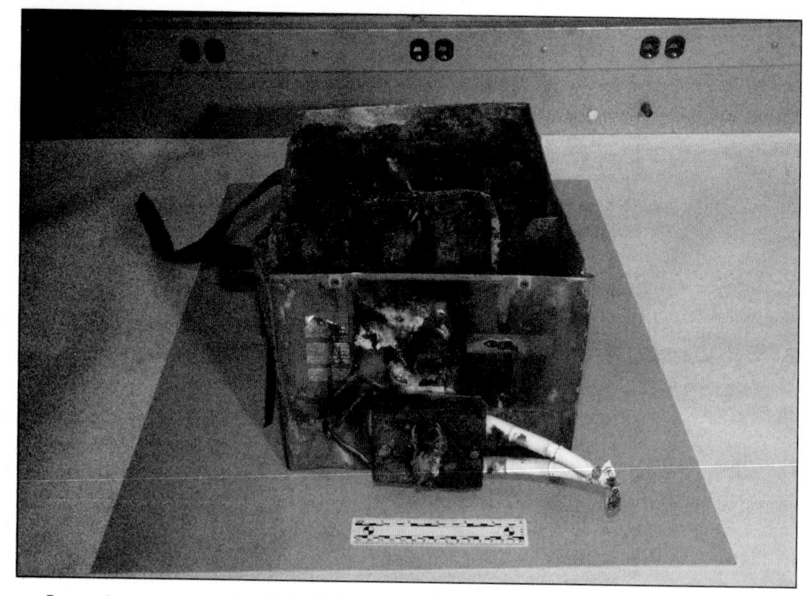

Investigators examined the lithium ion battery taken from a Japan Airlines 787 that appeared to have caught fire while the plane was on the ground at Boston's Logan International Airport on January 7, 2013. No passengers were on the plane at the time. | *(NTSB provided photo)*

NTSB investigator Mike Bauer examines the damage from the uncontrolled heating of a lithium-ion battery on a JAL 787. | *(Dreamliner NTSB photo)*

PART FOUR: HUMANITY

Capt. Dominic James was considered a hero after he successfully ditched a medical evacuation flight in the Pacific Ocean in 2009 with no loss of life. Later, top aviation officials in Australia vilified the pilot during an investigation that was itself subject to criticism. | *(Personal photo of Dominic James)*

Pilots failed to remove locking devices on the Boeing 299's flight-control surfaces, and the plane crashed during a demonstration flight in Dayton, Ohio, on October 30, 1935. The accident prompted the creation of the pilot checklist. | *(US Air Force photograph)*

The Pel-Air medical evacuation jet that ditched off the coast of Norfolk Island, Australia, photographed where it lay on the floor of the Pacific Ocean in 2009. | *(ATSB Photo)*

An artist's depiction of KLM captain Jacob van Zanten's attempted takeoff over the Pan Am 747 still taxiing on the runway. | *(Creative Commons)*

Adam Jiggins, a student at CTC Aviation, Hamilton, New Zealand, inspects his airplane under the supervision of flight instructor Sarah Jennings in 2011. | *(Photo courtesy the author)*

Lisanne Kippenberg during pilot training in Switzerland in 2013. | *(Kippenberg family photo)*

The first powered, controlled, sustained flight. Orville Wright is at the controls while Wilbur runs alongside. | *(Photo from Library of Congress)*

The author does a dismal job on a pilot aptitude test at CTC Aviation, a flight school for aspiring airline pilots in Hamilton, New Zealand. | *(Photo courtesy the author)*

PART FIVE: RESILIENCY

The nose gear of Air Canada Flight 143 collapsed as the 767 glided to an emergency landing in Winnipeg in 1983. The pilots did not know that a sports car club was holding a rally on the abandoned airfield but, miraculously, no one on the ground was injured. | *(Wayne Glowacki / Winnipeg Free Press)*

First Officer Laura Strand (*far left*) and Capt. Cort Tangeman (*far right*) with the crew of American Airlines Flight 1740 in Chicago. | *(Personal photo of Cort Tangeman used with permission)*

British Airways Flight 38 lands short of the runway after losing both engines on approach to Heathrow Airport. | *(Photo courtesy Metropolitian Police Air Support Unit)*

From left: Matt Hicks, Harry Wubben, Richard de Crespigny, Dave Evans, and Mark Johnson the morning after their brush with disaster in a severely crippled Airbus A380 over Singapore. None of the 459 aboard was injured in the emergency. | *(Personal photo of Richard de Crespigny)*

Nancy Bird Walton, the Airbus A380 flying as Qantas Flight 32, returns to Australia after repairs caused by an uncontained engine failure over Singapore in November 2010. | *(Photo courtesy Qantas)*

Snow Job

Of all the air accidents linked to political intrigue, the Air New Zealand Flight 901 disaster over Antarctica seems the most incredible. Featuring lies, manipulation, document shredding, and burglary and set in one of the world's most inhospitable places, it is a tale worthy of any tinfoil hat–wearing conspiracy theorist, except that it is true.

In 1977, long before cruise lines started offering tours of Antarctica, Qantas and Air New Zealand inaugurated one-day sightseeing trips through which adventure seekers could spend a day on a first-class aerial tour. Antarctica and back in just one day—who wouldn't want to take that flight? On Air New Zealand, the flights were scheduled during the near-twenty-four-hour days of the austral summer on a McDonnell Douglas wide-body DC-10.

From altitudes as low as fifteen hundred feet, passengers could see the Ross Ice Shelf; McMurdo Sound; New Zealand's scientific and industrial research center at Scott Base; the Americans' McMurdo Station, with its ice airfield; and even expedition sled dogs and indigenous emperor penguins. Most impressive was the still-active Erebus Volcano, the twelve-thousand-foot-high exclamation point of Ross Island.

These weekly trips were conducted only one month a year, so there was not much opportunity for pilots to gain experience in this unusual environment. Of the eleven flights between February 1977 and November 1979, only one captain had flown to Antarctica more than once.

It was considered a plum assignment. Management pilots handled most trips, often with VIP guests of Air New Zealand or journalists on board. So line pilot Capt. Jim Collins was surprised to be rostered on the flight on November 28, 1979. Flying with Collins would be first officers Greg Cassin and Graham Lucas and flight engineers Nick Moloney and Gordon Brooks. Brooks was the only one of the five who had flown an Air New Zealand Antarctic flight before. For the others, their first "Antarctic Experience" would be this one.

Shortly after arriving over the ice, the jumbo jet slammed into Mount Erebus, instantly killing all 257 people on board. At the time, it was unfathomable that the experienced airmen would descend into this well-known high terrain. Yet that shocking event was eclipsed by what happened next. The airline figured out how the plane happened to be over Mount Erebus; it was a terrible, tragic data entry error, as simple as it was unintentional. Rather than confess, however, the people at Air New Zealand decided to hide the truth by incriminating the pilots.

When Captain Collins made the decision to descend from sixteen thousand to six thousand feet to give his passengers a better view of Antarctica, he was confident that he was approaching the continent over McMurdo Sound, a forty-mile-wide expanse of water that ended at landfall at the McMurdo waypoint.

He and the two first officers thought this because it was the route they had been shown at a mandatory preflight briefing nineteen days earlier. It was also the route Collins and Cassin flew at the flight simulation session that followed, according to the testimony of two other pilots who attended the session with them.* The night before the crash, Captain Collins, a bit of a cartography nerd, sat in the dining room of his home in Auckland with his daughters and showed them the path he would fly, moving to the living room floor when one of the maps proved to be too big for the table. He pointed out to his two elder daughters, Kathryn, sixteen, and Elizabeth, fourteen, the mountains of Victoria Land, about twenty-five miles to the west of his overwater course, and Mount Erebus, twenty-seven miles to the east.

When Collins checked in at Auckland Airport the following day, however, the information given him to program the flight plan into the plane's inertial navigation system had been altered. In an attempt to correct a data entry error made fourteen months earlier, Air New Zealand changed by two degrees of latitude a waypoint on the routing. Rather than approach the continent over the flat terrain of McMurdo Sound, the plane would fly directly over Mount Erebus, the twelve-thousand-foot volcano. Even though this made for a very different flight path, the crew was not notified of the change.

Neither the captain nor the first officers would have noticed the difference just looking at the numbers, which are latitude

---

* First Officer Lucas did not attend the briefing.

and longitude coordinates for a series of waypoints between Auckland and the continent of Antarctica. The pilots would have had to pull out maps and cross-check the routing to catch the switch, which they had no reason to do. They had been briefed on the route and had practiced it in the simulator. That was the purpose of their November 9 session, and that is what Captain Collins had reviewed so intently the night before.

By the time the plane neared Antarctica, Flight 901 had been in the air for about four hours and forty-five minutes. This wasn't normal air travel. Air New Zealand wanted to delight and awe its customers, so it staffed the flight with an Antarctic expert, who provided a running commentary. Sir Edmund Hillary, who along with Tenzing Norgay was the first to summit Mount Everest, was on some flights. On this one, though, was Hillary's friend Peter Mulgrew, retired from the New Zealand Navy. Mulgrew had accompanied Hillary on an expedition to the South Pole in the late 1950s. Passengers moved around the cabin with glasses of champagne, looking out the windows and listening to Mulgrew pointing out the sights. They could mosey up to the cockpit, where an open door invited them to see the view head-on.

Cameras clicked as the plane passed the ragged mountains of Victoria Land to the west. In total, more than nine hundred images taken on the flight were developed and studied, photos that would play an important role in exposing flaws in the official conclusion of what happened to Flight 901.

Two hundred eighty miles north of McMurdo Station, Collins was invited by air traffic control at McMurdo Center (Mac Center) to fly down to fifteen hundred feet via radar, through a layer of clouds below him at eighteen thousand feet. This was good news: it meant that the passengers could get their first close-up look at the scene below. Then Collins saw the clouds breaking up, so he descended in clear air instead. On the way down, he made a modified figure eight. First he flew a three-hundred-sixty-degree turn to the west and then to the east, ending up once again aiming south toward McMurdo Station.

After completing the second orbit, Captain Collins reengaged the autonavigation system to be sure he was back on the course the airline had provided and continued his descent. Had the DC-10 been on the track shown at the pilots' briefing, this latest group of airborne explorers would have had the same experience as the travelers on earlier flights. But once Collins reactivated the automatic navigation, Flight 901 was locked onto a path that would take it straight into the side of Mount Erebus.

The cockpit and flight data recorders recovered at the crash site reveal that the plane was flying at around fifteen hundred feet when the first alert sounded with *Terrain pull up!* Seconds later, Captain Collins instructed the first officer, "Go-around power, please." The recording ends six seconds after the initial warning.

There were two big questions for those assigned to investigate the crash: why was the plane flying so low approaching high terrain, and why had the pilots not seen the mountain in front of them? For Air New Zealand, the answer to the first question came within hours.

As soon as they learned that the plane was missing, flight dispatchers Alan Dorday and David Greenwood looked at the navigation information given to Collins and compared it with the previous Antarctic flight. They saw that the two paths were different. Captain Collins was not routed over flat sea ice, as the previous flight had been, but directly over land, and not just land, but the highest terrain for miles around. The pilots were flying low over a mountain because they thought they were over the sea. It was that straightforward.

Anyone who has ever screwed up probably understands wanting to escape the consequences, and it must be exponentially greater when lives have been lost as a result. People at the airline gave in to that impulse. It began with a decision to hide the route shift from investigators so that when the country's chief air accident investigator, Ron Chippindale, left New Zealand for Antarctica the following day, he was unaware of the critical error that led the pilots to make their fatal descent.

Accompanying Chippindale for the ten days of on-site investigation was the chief pilot of Air New Zealand, Ian Gemmell. Unlike Chippindale, however, Captain Gemmell knew about the change in flight path, because flight dispatcher David Greenwood had told him.

Gemmell was by Chippindale's side throughout the investigation, as a technical adviser. How much he influenced Chippindale's early view of the accident is a matter of opinion. What is not disputed, however, is that materials discovered at the crash site were tampered with, and some things just went missing. The most intriguing item was Captain Collins's small ring binder—and it was the absence of this item that led people to suspect that Gemmell was behind the disappearance of evidence.

Forty-five-year-old former air force pilot Jim Collins was a notorious list maker and note taker. He took his ring binder with him in his flight bag whenever he flew.

As they worked at the crash scene recovering bodies, two New Zealand police identification officers, Stuart Leighton and Greg Gilpin, found the ring binder not far from where Captain

Collins's body lay. The binder had about thirty pages, mostly blank, except for a few in the front, which contained numbers. Sergeant Gilpin characterized it as flight-related information. When the binder arrived in New Zealand, however, those pages were gone. No one could satisfactorily explain what happened until 2012.

On his deathbed, Captain Gemmell told documentarian Charlotte Purdy that the airline had removed the pages. Purdy, whose uncle was the flight engineer on Flight 901, was producing the film *Operation Overdue*, an account of the retrieval of the victims from Antarctica. That someone whom Gemmell did not name removed the pages made sense to Purdy, but by then, all of New Zealand already knew about how the airline sought to cover up its mistake by keeping the route change secret and blaming the pilots for flying too low.

Yet there on the ice in 1979, any indication that Captain Collins did not know about the route over terrain would have derailed that plan. This is why the question of Gemmell's access at the crash site and his influence over Ron Chippindale was so important.

Gemmell always insisted he did not know about the alteration of the flight coordinates until after he returned from Antarctica on December 8, but his coworkers told a different story. "I certainly told Ian Gemmell," said David Greenwood, the dispatcher, when testifying about sharing his discovery with the chief pilot the morning after the crash.

These conflicting recollections did not happen in a vacuum, and it wasn't just Collins's binder pages that went missing. His atlas and flight papers were never found, either. And in the days following the accident, items were taken from the home of First Officer Cassin. Anne Cassin told me her husband

Greg's body had not even been recovered from Antarctica when she returned home from an errand to find that her in-laws had handed over boxes of documents to an Air New Zealand pilot who had come to the house. "Insurance details, Greg's time sheets, bills, receipts, personal letters, bank statements, check stubs, flying books—everything had gone," she said, the memory still vivid more than thirty years later. "The person chosen to liaise with the dead air crews' families and Air New Zealand stole every single item that I'll call paperwork from my home."

When she realized what had happened, Anne Cassin went to the coffee table where her husband had left the folder containing notes on his Antarctica briefing. Most of the contents were gone. Anne Cassin was a thirty-one-year-old mother of three and a private pilot,* but this early in the probe, she couldn't imagine why details of her husband's last briefing mattered to the airline. In February, it became clear when Air New Zealand's mistake in changing the route to McMurdo without notifying the crew was leaked to a newspaper.

Cassin and the pilots at Air New Zealand understood for the first time: The airline was saying the pilots knew they were flying over Mount Erebus. But the briefing notes indicated otherwise. That's why they were being snapped up. Collins's notes were gone and so were Cassin's. And all this time, Air New Zealand's chief executive Morrie Davis was ordering the shredding of crash-related documents.

That odd request came within days of the accident, when

---

* Anne Cassin went on to become a flight instructor and commercial airline pilot for Mount Cook Airline, in Christchurch.

the airline boss said that all original relevant material should be collected and everything else destroyed. Davis later explained to the lawyer for the airline pilots union that he was motivated by a desire to prevent leaks to the media.

In June 1980, Chippindale, the chief air accident investigator, released his report. The probable cause of the disaster was "the decision of the captain to continue the flight at low level toward an area of poor surface and horizon definition when the crew was not certain of their position." The airline was cited for the inaccuracies in the briefing to the crew. The country's Civil Aviation Division, or CAD, was told it should have paid more attention to how Air New Zealand was operating the flights. But it was the pilots who dominated Chippindale's report.

Regardless of whether the pilots thought they were flying over high or low terrain, everyone would have returned home safely if the crew hadn't descended below sixteen thousand feet, Chippindale said. There was a certain logic to this, reinforced by what the airline told Chippindale: pilots were absolutely prohibited from flying below sixteen thousand feet on the Antarctic Experience flights until directly over McMurdo.

It had taken Chippindale six months to come to his conclusions. One month later, a special Royal Commission Inquiry would begin examining all his findings and the information from the airline on which it was based. The national disaster was on course to become a national scandal.

Peter Mahon, a Christchurch lawyer and longtime judge in Auckland, was appointed to conduct the evaluation. After ten months, the commissioner found fault with practically every aspect of the probable cause report. It was full of inaccuracies and supported by sloppy fact-finding techniques, Mahon found. Curious interpretations of the truth were exposed in Chippindale's conclusions.

In one example, Chippindale wrote that officials with McDonnell Douglas and Air New Zealand pilots told him that when approaching Erebus, the pilots would have seen the volcano depicted on the flight deck radar monitor, so they must have ignored the approaching mountain. Yet Bendix, maker of the radar, said the air in Antarctica was too dry for the mountain to be reflected on radar. When Chippindale was asked to provide names or notes to back up his contrary information, he could not.

After a committee assigned to transcribe the cockpit voice recorder spent a week at the office of the NTSB in Washington, DC, debating every utterance and sound on the tape, Chippindale and Gemmell replayed the tape in Chippindale's home and revised the transcript. Chippindale then went to the United Kingdom and had another go at the transcript alone, revising it yet again.

"That's just not done. You cannot do that," aviation safety specialist John Cox told me. In his years as an accident investigator for ALPA, Cox has listened to and participated in the transcription of half a dozen cockpit voice recorders and calls them the most subjective aspect of the investigation. If Chippindale thought he'd heard something new or different in the cockpit voice recording, Cox says the process would have been to recall the entire CVR committee. "You sit down and say, 'We've listened again. We think the following corrections need to be made.'"

That did not happen. When the words were published in the final report, fifty-five changes had been made to what the CVR committee had earlier agreed to. Words and phrases no other contributor heard had been added, and an entire exchange the committee had agreed to had been excised.

The all-important question was whether the pilots had truly been ordered to remain at sixteen thousand feet on approach to the continent. Controllers in Antarctica seemed to have been unaware of the restriction, because they cleared Flight 901 to fifteen hundred feet. Mahon was convinced the prohibition was never heeded because flights often operated between fifteen hundred and three thousand feet. Newspaper reporters who made the trip wrote about the low-level flights, as did writers for the airline's marketing material.

The official probable cause report and the report of the Royal Commission disagree on just about every key fact except one: Air New Zealand and the Civil Aviation Division, which was responsible for regulating the airline, did not take seriously the risks associated with flying in Antarctica.

When the flights began, there were a number of requirements designed to mitigate the dangers. The crews were to be provided with flight charts on a topographical map. They were not. Two captains were to be on each flight, but that requirement was swapped for one captain and two first officers. All captains were to fly one flight with supervision. That rule was abandoned because the briefings were supposed to be good enough to eliminate the need for actual flight experience.

Above all, the special circumstance ignored by CAD and Air New Zealand is the one that sent Flight 901 into the mountain: the failure to recognize and train the pilots for the singular nature of flying in Antarctica.

People who knew Collins and Cassin to be thoughtful and conscientious aviators were outraged at the lopsided way the accident was attributed to the pilots. Still, everyone was baffled by one question: why steer a course into the side of a volcano? The day was clear: the photos captured in the cameras of the passengers revealed expansive landscapes, and sunshine to the east and west. Presumably, the view was the same out the cockpit windows, or Collins would not have continued to note that he was flying in visual conditions.

In spite of that, the official accident report concluded that the pilots were flying in cloud and that descent was foolhardy. The CAD's chief, E. T. Kippenberger, had a theory that "Captain Collins must have been suddenly afflicted by some medical or psychological malady, which made him oblivious to danger looming in front of him."

Ultimately, a more sensible answer, one in keeping with the evidence, came from Gordon Vette, an Air New Zealand pilot. Vette had been a captain on an Antarctic Experience flight. He'd flown with every crew member who died at Mount Erebus. He knew the pilots had been fooled into believing they were over McMurdo Sound. Reading the cockpit voice recorder transcript, he learned how the ground references on approach to Mount Erebus bore an unlucky and uncanny resemblance to what the crew would have seen on the McMurdo Sound approach. So he wondered if another trick had been able to hide from the crew a twelve-thousand-foot obstruction. Remarkable as it seems, the answer was yes.

When Collins took the plane below the layer of overcast, the summer sun was behind the plane. Ahead were forty miles of unbroken white: the white of the sea ice stretching to the horizon, where it blended indistinguishably with a sky bleached white by the sun's diffused rays above the unbroken clouds. The powerful effect of light meeting white eliminated the visual borders created by texture, shadow, and depth. Under these conditions, there would be no perception of the lines that separate one item of the landscape from another.

Dr. Arthur Ginsberg, director of the Aviation Vision Laboratory at Wright-Patterson Air Force Base, confirmed Vette's hunch that there was a reason for the pilots' improbable and inexplicable flight into the side of a volcano. The flat white carpet in front of the plane rising at an inclination of thirteen degrees and then nineteen degrees would not have been perceived by the crew. A pilot not familiar with the illusion would have flown straight into it.

Whiteout was not unknown at the time of Air New Zealand's Antarctic Experience flights. In his report, Chippindale devotes a page and a half to it. But whiteout (and more specifically, sector whiteout, where visibility is affected in just one direction) was not part of the training of the Antarctic crews.

Several movies and even more books have been produced dissecting the Erebus scandal. In *Verdict on Erebus*, his book about the case, Peter Mahon says that when he accepted the assignment, he thought Chippindale's report to the Office of Air Accidents Investigations was solid. He went in thinking he'd rubber-stamp a well-researched, well-reported analysis, but he discovered that nothing was as it seemed. What Mahon, an aviation outsider, saw, like a mountain dead ahead, was a government-owned airline that had engaged in what he ultimately called an "orchestrated litany of lies."

Mahon did not have the final word on Air New Zealand Flight 901. Chief Inspector Ron Chippindale wrote a rebuttal, claiming Mahon's finding "abounds in errors," and to this day, two conflicting reports account for one disaster.

How different is that really from the Arrow Air crash in

Canada, where neither probable cause was considered credible, or the Hammarskjöld accident, investigated four times and still considered incomplete? These troublesome investigations show that the search for truth does not always result in certainty, and that ambiguity may be the best cover-up of all.

When the Metrojet flight from Sharm El Sheikh in Egypt to St. Petersburg, Russia, crashed on Halloween 2015, confusion reigned once again. Metrojet executives immediately announced that nothing was wrong with the plane or the pilots, and British prime minister David Cameron offered that it was likely that a bomb brought down the airliner. Under international agreement, it fell to the Egyptians to lead the probe into what had happened, and for months they said it was too soon to say.

That didn't matter, because a deluge of stories concluded that terrorism was involved; even a front-page story in *The New York Times* said the Egyptians could not be trusted to carry out an impartial investigation.

A few days after the crash, I spoke to National Public Radio's Robert Siegel on the program *All Things Considered*. I told Siegel this should be no mystery: all the tin was there; the flight data and cockpit voice recorders had been recovered. What I failed to consider were the external factors, how each government would posture.

The Russians were responsible for supervising the safe operation of the airline. Had they failed to do so? European airlines were frequent users of the Sharm El Sheikh airport. Had they paid enough attention to the security there? Egypt had the most at stake; its economy is largely reliant on tourism. Had it been diligent enough protecting the safety of the tourists at its airports? Would it bamboozle, as the *Times* incautiously suggested, or were its investigators acting responsibly by holding off on drawing conclusions?

As much as metal on the ground, these intangible elements of realpolitik become part of the story, and the growth of international air travel means that even more than in the past, an accident investigation will involve multiple governments. Their competing interests are thought to provide balance and to ensure an unbiased result. But what I say to Susan Williams, I say to you: these cases caution us to be judicious with our optimism.

PART THREE

Fallibility

It's slips, lapses, mistakes, and violations. It's tedious because it is so banal. It's like every day, it's like breathing, it's like dying. There's nothing remarkable about it.

—PSYCHOLOGY PROFESSOR JAMES REASON

Progress and Unexpected Consequences

In the summer of 2011, I spent a week in Dubai with a hundred enthusiastic twenty-somethings training to be flight attendants for Emirates Airline. I was writing a first-person article about how Emirates is changing the industry by bringing glamour back to air travel.

At the end of my stay, I flew back to New York on an economy ticket, but I didn't stay in the back of the plane for long. Recognizing a pin I was wearing that indicated I had attended a secret initiation ceremony for Emirates employees, a curious flight attendant moved me to an empty spot in business class. I was digging into a bowl of salted almonds and trying not to get too much grease on the controls of my wide-screen TV when the woman came back with even better news. I was being moved yet again, now into first class, with an even cushier seat and a bigger TV screen. I wasn't there long before the spa attendant—yep, you read that right—came to schedule my appointment to bathe in what was, at the time, the world's only commercial airline in-flight showers. Frankly, it didn't sound especially appealing, but I reasoned that such an opportunity would not likely come my way a second time.

Standing in the altogether in the narrow cylinder watching the timer-activated faucet counting down my five-minute water limit, I marveled at how much had changed in air travel. And you should, too.

The Airbus A380 on which I was flying takes off weighing as much as 1.3 million pounds. It carries 555 people in three classes, or as many as 853 in the all-economy version. It is a far, far cry from the first airliners, like the canvas-covered Avro 10 Fokker with wicker seats for eight passengers that was flown by Australia National Airways in the early '30s, or the Martin 130 flying boat with sleeping berths for the long transpacific routes flown by Pan Am just a few years later.

The twenty-first-century airliner soars seven miles above the earth, a warm bubble protecting travelers from the dry, frigid, thin troposphere just beyond the cabin walls. The French-built, shower-equipped Airbus A380; the equally revolutionary American-built Boeing 787 Dreamliner, which is so sophisticated it is called a computer network with wings; Canada's new Bombardier CSeries; and Brazil's growing menu of Embraer E-Jets, demonstrate what the best engineering minds can produce. Each new model adds to the knowledge base. Even when the engineers err, their mistakes become bricks in the foundation for the new and improved model.

Err is such a tidy word, suggesting manageability. But to err when you are an aeronautical engineer is to unleash mayhem. Planes that do not stop upon landing, that issue alerts that confuse pilots, that explode and catch fire or pitch down unexpectedly—these are the unanticipated products of progress in aviation, and it has been that way since Wilbur Wright told the Western Society of Engineers in Chicago in 1901, "If you are looking for perfect safety, you will do well to sit on a fence and watch the birds; but if you really wish to learn, you must mount a machine and become acquainted with its tricks by actual trial."

Those tricks beset the Wrights on September 17, 1908. During a demonstration flight of the Wright Flyer for the U.S. Army at Fort Myer, Virginia, Orville Wright lost control of the plane, and it hit the ground. He was injured, and his passenger, Lt. Thomas E. Selfridge, was killed. It didn't take long to figure out what happened: A propeller blade had split, hitting a bracing wire. That, in turn, tugged on the rudder, nosing the plane down. As planes became more sophisticated, the factors leading to catastrophe became more numerous and more difficult to diagnose.

No example is more cited than the fatal design defects built into the world's first jetliner, the de Havilland Comet. The British-designed mid-twentieth-century airplane wasn't just the first passenger jet; it was a spectacularly innovative design with four motors tucked smartly into the wings. Even today it looks futuristic.

The story of the Comet focuses on its propensity to burst apart in flight, its most dramatic and well-publicized flaw, but it had other design issues. Accidents prompted changes, and lessons were learned. What has not changed is how difficult it is for designers to know in advance all the ways an idea on the drawing board will function in reality.

This is not surprising. To create an airplane is complex, involving layers of decisions that become locked into systems that are inextricable parts of the whole. Undoing one feature is like trying to remove eggs after they have been beaten into cake batter.

In the case of the Comet, a number of problems emerged

soon after passengers started flying on it. On two occasions the plane failed to take off, careening instead off the end of the runway. Three more events were even more mysterious: the planes simply broke up in the sky.

De Havilland may have been the first jet maker to go back to the drawing board, but it was far from the last. The flap handle on the DC-8, the cargo door on the DC-10, the fuel tanks of nearly all Boeing airliners—these and other components have been reexamined, reconfigured, or redesigned.

In what must be one of the fastest redos in history, Boeing modified the battery system on its 787 Dreamliner after the worldwide grounding of the fleet for nearly four months in 2013. Relentless news coverage about the things Boeing overlooked prompted the chief engineer for the 787 Dreamliner, Mike Sinnett, to admit that revolutionary creations are never fully understood at the outset. "Unknown unknowns," as he called them, lurk within, and the process of transitioning to the known is a messy and sometimes unpleasant affair.

Nearly every mishap-induced airliner redesign features a Jeremiah, the early detector of the "unknown unknown" who voices concern about the problem, but who may or may not be heeded. For the Comet, the first Jeremiah was Capt. Harry Foote in 1952.

The thirty-six-year-old pilot flew for British Overseas Airways Corporation (BOAC).* He worked with aviation writer and fellow BOAC captain, the late David Beaty. In his book *Strange Encounters: Mysteries of the Air,* Beaty concludes that Captain Foote's suicide at the age of fifty-three was because of his role as the first pilot to crash the world's first jet-powered airliner.

---

* Now known as British Airways.

The crash happened just six months after the Comet's first flight with fare-paying passengers, on May 2, 1952. That historic event was followed just a few weeks later by the first Comet flight with royalty on board. Queen Elizabeth; her sister, Princess Margaret; and the Queen Mother had a four-hour fly-around as guests of de Havilland.

All this proclaimed that the Comet wasn't just a new airplane; it was the vehicle that was going to fly England into the future. The postwar nation was getting back on its feet and taking to the air. British products were flying into the global marketplace.

And why not? Jet technology originated in Britain, the invention of a young Royal Air Force cadet named Frank Whittle, who patented the idea for a gas turbine engine while still in his twenties. And though the Germans were the first to create a jet that actually powered an aircraft, Whittle went on to help develop the engine used for the DH-100 Vampire, England's first single-engine jet fighter. With this technological advantage, the British could overtake the American plane builders Lockheed, Douglas Aircraft, and Boeing, who were selling their slow and noisy propeller planes to airlines around the world.

In 1947 the British government turned to the Vampire's creator, Geoffrey de Havilland, to make the first jetliner. It took less than a year for the company to detail the new plane's attributes, and they were astonishing. The airplane would be light, with a thin aluminum skin. Some sections would be glued rather than riveted, saving the weight of metal connectors. And the plane would fly high. Where the DC-6s and DC-7s and the Lockheed Constellations cruised at twenty-four to twenty-eight thousand feet, the Comet would soar at thirty-six to forty thousand feet. In the thinner air of higher altitude, the plane would encounter less drag and would be more fuel efficient.

The downside of flying seven miles above the earth was that the cabin would require an unprecedented level of pressurization. An interior atmosphere equal to about eight thousand feet above sea level meant putting eight and a half pounds of pressure on every square inch of the walls separating outside from in.

More pressure and a thinner structure were two decisions that would factor in the disasters soon to come, but they were not the only ones. Those new jet engines would subtly alter some basic characteristics of flight and have their own repercussions.

When the war ended, military aviators filled the cockpits of BOAC's airliners. Among them was Harry Foote, who flew the four-engine Lancaster heavy bomber for the Royal Air Force. He had 5,868 hours in his logbook and was considered one of the airline's elite pilots, according to Beaty, who wrote that only the best were selected to fly the technological marvel that was the Comet 1. Of course, even the elite would have relatively few hours in the new plane. Foote had just 245 on the Comet on October 26, 1952, the day that would mark the beginning of the end for Harry Foote and for the Comet 1.

Unknown Unknowns

It was just before 6:00 p.m. on October 26, 1952, the sun had already set, and rain was falling when Captain Foote began the Comet's takeoff roll for the second leg of a journey from London to Johannesburg. Eight crew and thirty-five passengers were on board the plane, registered as G-ALYZ. As the plane accelerated on the runway at Rome's Ciampino airport, the pilots watched for the needle on the speed gauge to reach eighty knots. When it did, Foote pulled back on the yoke and felt the nose of the airplane rise. The main landing gear was still on the ground, as it should have been, as the plane continued to accelerate. At one hundred twenty knots, Foote pulled the control column back again to lift the jetliner into the air.

So far, it felt like every other takeoff, so he called for the next step. "Undercarriage up," he said to the first officer. Before the man had time to comply, however, the left wing dropped and the plane turned left. The plane was no longer gaining speed, and the pilots felt a buffeting sensation, the precursor to a stall. The Comet flopped back onto the runway, forward momentum now propelling it into the darkness.

Foote pulled back on the throttle, cutting fuel to the engines as quickly as he could, but what had caused the plane to slow was the braking effect of the main landing gear being torn away by a mound of earth. The plane stopped just ten yards from the perimeter fence. Mercifully, there was no fire, though one wing had ruptured, spilling fuel onto the ground and sending fumes into the night. Passengers were shaken but uninjured.

By November, an accident investigation concluded that Foote's technique, lifting the nose too high on rotation, was the cause of the crash. The pilot argued that it hadn't happened that way. The airplane had become airborne after rotation, but had failed to climb and instead sank onto the ground, nose still high.

Many forces were working against Foote. The first crash of the highly touted Comet was the kind of news money can't bury. Banner headlines and news photos showed the gearless, semi-wingless, entirely hapless G-ALYZ as it lay on the far end of Ciampino airport. Even those at BOAC who sympathized with Foote's argument would not help him in his effort to reopen the examination, which had acquitted the plane in less than a month by convicting the pilot.

Then, just four months later, Capt. Charles Pentland of Canadian Pacific Airlines had a similar problem getting the plane off the ground. What started out as an uneventful takeoff roll from the airport in Karachi, Pakistan, ended in a deadly inferno. The plane was not carrying passengers, only four crew members who worked for the airline and six technicians employed by de Havilland, all of whom were killed.

Canadian Pacific was taking delivery of the plane it had

already named *Empress of Hawaii*, because it was going to provide the Sydney-to-Honolulu leg of the airline's transpacific service to Vancouver. The *Empress* was making a hopscotch journey. Day one was London to Karachi. The second day began at 3:00 a.m. On an already hot and steamy morning, the crew prepared for the flight from Pakistan to Rangoon, Burma.*

More than half a century later and after decades of research into how to enhance pilot performance, it is apparent how many aspects of the *Empress* flight created additional hazards for the pilots. Most significant, neither of the two men flying the *Empress* had even the little experience flying jets that BOAC's Comet pilots had.

Captain Pentland, the airline's manager of overseas operations, and Capt. North Sawle, thirty-nine, each had thousands of hours of flight time, but the Comet was their first experience in a jetliner.

In his book *Bush Pilot with a Briefcase*, the biography of Canadian Pacific's then-president Grant McConachie, author Ronald Keith describes Pentland's and Sawle's Comet training as a "crash course." The terrible pun notwithstanding, even the instructors at the de Havilland pilot school in Hatfield, England, considered the men novices during their time there.

Ironically, part of what made them weak in the Comet was the depth of their experience on other aircraft. Pentland had been a pilot with BOAC and Imperial Airways before joining CPA. Captain Sawle was CPA's chief pilot for overseas operations. He had been an aircraft mechanic, a plane builder, and a float and ski plane pilot who in his youth had flown mail,

* Now Yangon, Myanmar.

supplies, and passengers around some of the least hospitable areas of Canada's frozen north.

What Pentland and Sawle learned at Hatfield "clashed with flying instincts formed by many thousands of hours at the controls of conventional planes," according to Keith. On the night of the crash that killed them, Keith wrote that "neither had experienced a night take-off in the jet, nor had they flown it heavily loaded."

Yet the company not only selected these two to fly an unfamiliar plane on a globe-spanning delivery flight without any relief pilots, but McConachie had ratcheted up the pressure by trying to set a publicity-generating London-to-Sydney speed record. It was a decision Pentland called "bloody rough on us cockpit help."

Aviation at the time was a swashbuckling, take-no-prisoners business, with airline bosses such as Canadian Pacific's McConachie, Pan Am's Juan Trippe, TWA's Howard Hughes, and American's C. R. Smith. These men presided over equally driven superpilots who claimed to thrive on the knife's edge, impervious to the fatigue and fear that affected ordinary men. It was an unrealistic dynamic, ridden with risk. Suffice it to say, Pentland and Sawle were set up to fail even if the plane had not harbored the many design flaws it did.

Loaded with two tons of fuel in the wing tanks, *Empress of Hawaii* was taxied into position by Captain Pentland and prepared for takeoff. Pentland set the brakes and advanced the throttles, feeding fuel to the four engines. He watched the gauges, waiting for the engines to gain sufficient power so that when he released the brakes, the heavy plane would pop off the mark and begin the takeoff roll.

Moving down the runway, the plane passed eighty-five knots. Pentland pulled back on the yoke, raising the nose of the airplane. He expected the acceleration to continue, but the plane instead lumbered on at one hundred knots, twenty-two knots below what was required to get it into the air.

Whether because of fatigue or habit, Pentland seemed to have forgotten that takeoff in the Comet requires a different procedure. At rotation, the nose is elevated slightly, just three to six degrees: anything steeper risks making the airplane stall, even while it is still on the ground.

The plane had already consumed more than a half a mile of runway and was still far from achieving takeoff speed. Concerned, Pentland raised the nose even higher, but the plane went no faster. At that moment something must have clicked. Pentland lowered the nose, and the wheel hit the runway. Only then did the plane start to speed up. He needed just a few seconds of acceleration now, but those were seconds he did not have.

The plane was going too slow to fly and too fast to stop when the Comet reached the end of the runway. The right landing gear hit a culvert, causing the plane to pivot sideways into a dry canal and then slam against the forty-foot embankment on the other side. The brand-new Comet shattered into pieces and erupted into flames as two tons of fuel burst out of the broken tanks in an explosive mist.

This second event, now with fatalities, exposed to a wide audience the first of the Comet's "unknown unknowns." Harry Foote was aware that something was wrong from his firsthand experience of four months earlier. The official report may have blamed his piloting skills, but it was notable that de

Havilland revised the takeoff procedures for the airplane after his event.

The new procedures required the pilot to rotate at eighty knots and then lower the nose back to the ground while increasing speed. That was the takeoff procedure taught to Pentland and Sawle when they arrived for Comet training early in 1953.

In an article for *American Aviation*, William Perreault and Anthony Vandyk explained that unlike with piston planes, where the propellers "create a bubble of compressed air close to the ground that nudges the wing up," the Comet wing with its jet engines did not get the benefit of this cushion. The engineers at de Havilland had discovered "a downward push which can make it stall," the article said. Pilots were instructed to nurse the plane off the ground by raising, lowering, and then slightly raising the nose again at takeoff speed, a technique called the "Foote takeoff."

Despite these two accidents, the Comet was flying high among the traveling public, on a steady course to three more disasters and an aviation mystery that would command the world's attention.

While the Comet 1's propensity to ground-stall was unrelated to the design problem that would bring the plane to infamy, Foote's personal experience on the wrong side of the first design error turned him into the Comet's fiercest critic. He could never put his finger on what was wrong with the airplane, partly because the only other pilot who he knew had experienced the ground stall on takeoff was now dead. Foote was unaware of any other similar events. During a series of meetings with the British Air Line Pilots Association, in an attempt to reopen the investigation that had tarnished his reputation, Foote learned that eight other BALPA Comet pilots had been disciplined for their involvement in accidents during the brief but deadly year and a half the Comet 1 had been flying. He could also see that de Havilland was making changes associated with the ground-stall issue. The takeoff procedure had been revised a second time. Later versions of the airplane featured stall warning devices and a redesigned wing that provided more lift.

Foote's thoughts kept coming back to the Comet because events related to the plane were finding him like the proverbial bad penny. After his accident, BOAC had sent Foote back to flying propeller planes. He considered this a demotion, but he accepted it. A few months after his accident, Foote was visiting the training center at Hatfield; Captain Pentland was also there, for his Comet training. The two shared a car to the train station, but not a conversation. Foote told fellow BOAC pilot David Beaty that it was an uncomfortable ride.

Imagine the scene: two captains, one sanguine about his upcoming status as jetliner captain, the other humiliated for the flaws in that same plane and now sentenced to spend the rest of his career flying yesterday's piston-driven aircraft. It was a two-character Greek tragedy set in the backseat of a car.

Later, when Foote heard the news about the accident that killed Pentland, he shared his worries with Beaty, suggesting that if anyone had paid attention to what he had been saying about the odd takeoff characteristics of the plane, Pentland and the ten others on the *Empress* might still be alive. His argument was getting traction, because shortly after that, *Aeroplane* magazine made a similar point. The article said that after Foote's crash, the chance of "pilot error could be accepted" as one in a million, but with the *Empress* crash just four months later, the plane's design must be considered. When pilots repeatedly make the same mistake, it "must be presumed to be too easily possible."

On May 2, 1953, the first anniversary of the Comet's inaugural flight, Foote was in command of a four-engine York freighter headed to Calcutta. In the sky twenty thousand feet above him, a BOAC Comet was flying in the same direction. Foote's plane was carrying cargo; on the jet there were happy passengers being waited on by an attentive staff. After a stop in Calcutta, many of the travelers would continue on to London; several were going to the coronation of Queen Elizabeth, on June 2.*

Given the jet's speed advantage, the Comet landed to refuel for its journey to Delhi well ahead of Foote's lumbering York, but the two pilots met up at the airport. While the captain of the Comet, Maurice Haddon, waited for late-arriving passengers to board, he and Foote chatted briefly and then went their separate ways, Foote to his hotel, Haddon back to the flight deck of the airplane with the registration G-ALYV.

When the last of the Comet's passengers was seated, the plane departed. After takeoff, Captain Haddon confirmed clearance to climb to thirty-two thousand feet. That was the last message transmitted. Arriving at the hotel, Foote heard the news that the Comet had gone silent. BOAC tried to allay any fears. "We have not posted it as missing yet," a spokesman for the airline told the *Times of India* news service. "We hope that the [air]liner has come down at one of the emergency landing

---

* Queen Elizabeth had already taken the throne, but her coronation was delayed to observe a period of mourning for the death of her father, King George VI, in February 1952.

grounds between Calcutta and Delhi." Foote was not as optimistic.

At dawn the following day, his was the lone British airplane among those from the Indian Air Force taking off in search of G-ALYV. Twenty-five miles to the northwest, Foote spied the wreckage in an area accessible only by foot. So widely strewn was the debris that it appeared the plane had disintegrated while still in the air. According to his friend Beaty, when Foote heard of the plane's disappearance the night before, he was heard to have predicted, "It will have crashed."

The area where the plane went down may have been remote, but it was far from unpopulated. During an evening of sixty-mile-an-hour winds accompanied by dust storms and torrential rains, villagers reported explosions and flashes of light. One boy said he'd seen a wingless plane flying low. Where the pieces of the Comet hit the ground, a man told the *Times of India* news service that he heard human screams amid the fire, "but the heat was so intense that no one could approach."

The storm was immediately thought to have been the cause of the crash. A policeman in the town of Jangipara telegraphed the news to Calcutta, wiring, PLANE KNOCKED DOWN BY TEMPEST. When investigators began their work, they found both outer wings separated from the plane, explaining the "wingless machine" seen by the villagers. In the report the government issued a month later, the Indian inspector of accidents was circumspect, explaining that his panel had suffered from limited facilities and data and a lack of time to investigate adequately. A year or more would be needed, the report said. But they were clear about one thing: "the aircraft suffered a complete structural failure in the air" during a thunder squall. In a joint statement, BOAC and de Havilland challenged the Indians' conclusion that this was because of either "severe gusts or over controlling or loss of control by the pilot" as little more than theorizing.

To know what really happened, the Royal Aircraft Establishment, the British government's research-and-development agency, needed to determine the sequence of the breakup. While the official cause of the loss of the first two BOAC Comets had faulted the pilot, this time pilot action was an unlikely cause, the de Havilland/BOAC statement said. If it was an attempt to protect the reputation of the airline's pilot training and the integrity of a jetliner that had already been sold to airlines around the world, the effect was lopsided. Airlines were not reassured; orders from Japan, Brazil, and Venezuela were canceled. Among passengers, however, the plane remained popular. BOAC had no problem selling tickets to fly on the Comet.

Even before the Calcutta crash, de Havilland had been conducting additional research to determine if the plane was experiencing metal fatigue. The company had been prompted to do so based on what it was learning about age and fatigue on de Havilland transport planes used by the Royal Air Force. Initially, the primary focus was on the wings, but soon the U.K. Ministry of Supply, which oversaw the military planes, and the Air Registration Board, responsible for civil airplanes, asked de Havilland to broaden the study. De Havilland's chief structural engineer, Robert H. T. Harper, agreed.

The work began in July, with technicians repeatedly applying levels of pressure to the cabin greater than what it would experience in flight. By September, the engineers found that tiny cracks were developing in the aircraft skin at the corner of the plane's square windows.

This sounds like an aha moment, but hang on. Rather than causing alarm, the examiners found it reassuring. The amount of pressure applied to the cabin walls during the testing was so far beyond what the plane would experience in flight, "It was regarded as establishing the safety of the Comet's cabin with an ample margin."*

What de Havilland failed to realize was that the test setup provided additional support to the structure that would not be there in flight. This was a critical oversight, one that led investigators to have a false appreciation of the cabin's strength under pressure. This same kind of error would be made sixty years later, during the testing of the revolutionary Boeing 787 Dreamliner.

Pressure tests were also conducted on the wings of the Comet, and in December, in much the same way, tiny cracks began to appear. In this case, however, the engineers were alarmed. The pressure had been applied for the equivalent of six thousand hours, and that was not a lot of flight time. Many Comets had already flown more than that. BOAC immediately instituted an inspection program on its planes.

Throughout Christmas, the manufacturer and the airline discussed whether that was enough. Did the wings need to be modified? The debate continued until it was interrupted with another Comet bursting apart in the sky.

---

* "Report of the Public Inquiry into the Causes and Circumstances of the Accident Which Occurred on the 10th January 1954, to the Comet Aircraft G-ALYP, http://lessonslearned.faa.gov/Comet1/G-ALYP_Report.pdf."

This was the original plane—the first jetliner to carry passengers, the plane that had starred in the newsreel films. Now it was in pieces on the floor of the sea and thirty-five people were dead.

It happened on January 10, 1954, a clear, sunny Sunday. Capt. Alan Gibson, thirty-one, and thirty-three-year-old First Officer William John Bury had an uneventful takeoff from Rome for the last leg of a flight that originated in Karachi. Rome to London would have been a quick two hours and twenty minutes in the air.

Captain Gibson was ascending through twenty-six thousand feet and making a radio call to a BOAC pilot in another plane, but the transmission was cut off midsentence. At thirty thousand feet, the center of the passenger cabin ripped open. The tail, the nose, and the wings outward of the embedded engines blasted off with a downward force. The tail remained largely in one piece as the plane tumbled through the air and slammed open-end first into the water.

Before the rupture of the passenger cabin, travelers had been seated two by two on either side of the plane's center aisle in armchair-style upholstered seats that might have made them feel as if they were enjoying a cocktail in the living room of friends. With the breakup of the airplane, some of those travelers were shot out of the ruptured fuselage. Fifteen others slammed against what remained of the fractured front bulkhead, which was fitted with a library shelf and water dispenser. The torn wings exposed the engines and released flammable kerosene that quickly ignited. The fire spread inward, burning the bodies of those who remained in the cabin.

As in the in-flight breakup of G-ALYV outside Calcutta,

G-ALYP shattered into countless pieces, descending in a cacophony of explosions and smoke. In India, the witnesses had been jungle villagers. In Italy, they were fishermen who watched for three long minutes as the plane and its contents fell into the water near the Tuscan island of Elba. Racing out in their boats, they came across the only bodies that would be recovered, the fifteen trapped on board by the bulkhead.

Autopsies by the director of Pisa's Institute of Forensic Medicine, Dr. Folco Domenici, would provide important clues about what happened. The victims' organs showed the plane had experienced a split-second decompression at high altitude and blunt force injuries suggested they had died from slamming into the cabin divider.

It took awhile for investigators to recognize the value of that information, but its most immediate effect may have been to provide a shred of comfort to the victims' families: their loved ones could not have suffered. They died immediately.

Floating wreckage was gathered, including the rear fuselage, the engines, and the wing center section, but the rest of the plane lay six hundred feet below the surface of the Tyrrhenian Sea. At that depth it was twice what British salvage divers could descend to at the time.* Obtaining it would not be easy, and certainly it would not be quick.

Investigators from BOAC, de Havilland, and the Air Accidents Investigation Branch of the U.K. Ministry of Transportation focused instead on the airplanes to which they had access.

---

* Tony Booth, *Admiralty Salvage in Peace and War 1906–2006: "Grope, Grub and Tremble"* (South Yorkshire, U.K.: Pen and Sword Maritime, 2007).

All seven BOAC Comets were grounded in what the airline called a prudent measure "to enable a minute and unhurried technical examination of every aircraft in the Comet fleet."

De Havilland threw itself behind newer, safer versions of the airplane. The engineers examined the structures and the systems. They discussed all the possible scenarios. They designed modifications to guard against anything that could have happened, even if no proof existed that the potential problem had anything to do with the disaster. If the investigators found something that didn't happen but could have, it became a high-priority fix. Note that, because you don't see it happening every day.

Lord Brabazon, the chairman of the Air Registration Board and a member of the Air Safety Board, summed up the work by saying, "Modifications are being embodied to cover every possibility that imagination has suggested as a likely cause of the disaster. When these modifications are completed and have been satisfactorily flight tested, the Board sees no reason why passenger services should not be resumed." So even though the cause of the crashes remained a mystery, on March 23 the Comets returned to the sky.

Of the seven airplanes subjected to the alterations described by Lord Brabazon, one was a two-and-a-half-year-old Comet, registration G-ALYY, with twenty-seven hundred flight hours. In February, the plane had undergone an eleven-pounds-per-square-inch pressurization test to check its structural soundness. G-ALYY was leased to and operated by South African Airways to service the London-to-Johannesburg route. On April 2 and again on April 7, the plane was subjected to more inspections, perhaps one too many.

After a panel was removed from the plane to give access to a test inspector, the panel was not reinstalled correctly. The plane departed on its first leg to Johannesburg on April 7. On arrival in Rome, the mechanics were horrified to discover loose bolts inside the right wing and "an equal number of missing bolts" from the panel at the wheel well. Opening the plane up for inspection had created a hazard. That nothing had gone wrong as a result seemed a blessing, but the blessing was short-lived.

Other maintenance issues kept the plane in Rome for a day. At 6:32 the following evening, former South African military pilots Capt. Willem Karel Mostert, thirty-eight, and thirty-two-year-old First Officer Barent Jacobus Grove, took the plane and its fourteen passengers into the air. Next stop: Cairo, three hours away.

Radio calls were normal. The crew checked in at seven thousand feet, and then at eleven thousand feet. Then, on the way to thirty-five thousand, they made a transmission that also included telling controllers their anticipated arrival time in Egypt. Then nothing more was heard. Calls from Rome and Cairo went unanswered.

The scenario was all too familiar.

It took a day just to find the oil slick indicating that the flight had likely ended in the Tyrrhenian Sea off the coast of the volcanic island of Stromboli. Six bodies were found along with some airline seats, but in an area where the sea depth was thirty-three hundred feet, the downed plane was as inaccessible as if it had flown into outer space. Concluding that the wreckage would never be recovered, authorities realized that whatever they determined about this accident would be an extrapolation from what they found in the Elba crash.

Both planes had come apart on ascent at roughly the same altitude and with tremendous force. The plane that crashed at

Elba had flown 1,290 pressurized flights; the South African Airways plane had completed 900. Again, the autopsies were performed by Dr. Domenici, who confirmed that a rapid decompression had taken place. The question was why.

The British government pulled the airworthiness certificate of the Comet 1. This time the planes were on the ground for good.

Into this puzzle came one of the era's most provocative thinkers, Alan Turing, who broke Germany's Enigma code and was the subject of the 2014 movie *The Imitation Game*. Turing developed the Automatic Computing Engine, a machine that automated complex equations so that they could be completed faster than humans could solve them. Parts of G-ALYP retrieved from the sea were subjected to exhaustive testing and comparison to an undamaged Comet, and for that, Turing's Pilot ACE computer was used to run the many calculations required.

The investigators didn't rely entirely on that newfangled thing called a computer. At the Royal Aircraft Establishment in Farnborough, workers also constructed an enormous tank into which would fit the fuselage of an entire Comet 1 jetliner, with the wings protruding out on either side like arms out of an undershirt. Beginning in early June 1956, water was alternately pumped into and out of the cabin to create 8.25 pounds of pressure per square inch against the walls. This was to simulate the effect of flight at forty thousand feet. Each infusion and release of water lasted about three hours, a typical flight leg for a Comet. Every day, the test airplane accumulated virtual flights and experienced actual stress.

While that was happening, the wings were flexed up and down as if the plane were in flight. After one thousand applications, a "proving flight" was conducted and the pressure to the inside of the plane was increased to eleven pounds per square inch.

On June 24, as the water was being pumped into the plane for the proving flight, the needles of the pressure gauge passed 8. They climbed to 9 and then 10. But it got no farther than 10.4 inches of pressure, because the fuselage ruptured at the cabin ceiling. An eight-foot section gaped open as much as three feet at the widest part. The slice ran through an area that included the cutout for the emergency escape hatch.

That might sound like the end of the story, but was far from it. It was as if a light had been turned on a previously darkened path. Investigators could see where they needed to go, but not what they would find on arrival. The effort to retrieve wreckage from G-ALYP off the coast of Elba received new energy. Two months later, nearly 70 percent of it had been recovered, enough to show that there were several places where a fatigue crack could have originated.

It was not possible "to establish with certainty the point at which the disruption of the skin first began," the report of the Comet inquiry read, but the long, erratic, and oftentimes faint line charting the course of this mystery was soon to end.

When the Elba accident investigation was concluded, it was joined like a Siamese twin to the tragedy that followed four months later near Stromboli. The report lays out the whole sad story, justifying to the end the decisions by British aviation officials to keep the Comet in the air. Many, including me, would wonder about that. But when I ask those who have studied the history of the Comet, I find few skeptics.

Of the final decision to let the plane fly after the still-unsolved Elba accident, Graham Simons, aviation historian and author of *Comet! The World's First Jetliner*, writes, "It was all that could be done, for no one had any idea what had gone wrong" with the airplane. Hold on to that thought, because Simons has summarized not just the decision to let the Comet fly again, but the kind of risk-benefit analyses airplane manufacturers have been making ever since.

Deflection

The single aisle Boeing 737, introduced in 1968, is Boeing's best-selling airliner and has flown in nine different versions. On March 3, 1991, a United Airlines 737 crashed preparing to land in Colorado Springs, killing all twenty-five people aboard. It had been an uneventful flight until just after 9:43 a.m. Flight 585 was on approach, and First Officer Patricia Eidson was concerned about the turbulence experienced by the plane that had landed ahead of them.

"I'll watch that airspeed gauge like it's my mom's last minute," Eidson said to Capt. Hal Green. Indeed, the descent was dodgy, with the plane accelerating and Green complaining how hard it was to hold his airspeed.

"Wow," Eidson said, followed twenty seconds later by "We're at a thousand feet." The plane rolled to the right, and the pilots tried to regain control. The cockpit voice recorder suggests the captain was adjusting settings for a go-around, abandoning the present landing to go around and try again. The only communication between the pilots makes it clear they realized the plane was going to crash. "Oh God," the first officer says repeatedly. "Oh no," Green says the second before impact.

Investigators thought early on that the rudder, the hinged vertical panel on the tail that swings from side to side to control the plane's left and right motion, played a role. The week before the accident, two crews who flew the plane reported problems related to the operation of the rudder, including uncommanded movement. In July 1992, as the investigation was under way, a United maintenance worker reported finding an anomaly during a ground check of another 737 in the fleet. "The rudder had the potential to operate in a direction opposite to that commanded by the flight crew," the airline reported. The main rudder control valve was changed by Parker Hannifin, the company that designed it.*

Still, after nearly two years studying the evidence, Tom Haueter, the NTSB investigator in charge, could not say conclusively if or how the rudder factored in. The probable cause of the accident was left as one of two likely events: some mechanical problem with lateral control of the aircraft, or an atmospheric disturbance that caused the airplane to enter an uncontrollable roll. From that point on, though, Haueter took notice of any difficulties reported with the 737's rudder.

Among the family of Boeing airplanes, the 737's rudder design was unique in a number of ways. First, since it was smaller than the 727 and the 747, there was not enough room for two entirely separate and redundant power control units. That unit, known by the abbreviation PCU, takes the input of the

* This was the first of four modifications to the Parker Hannifin design.

pilot's foot on the rudder pedal and converts it through hydraulic action into movement of the swinging panel on the tail.

Boeing got the plane certified with a novel design that put both control of the rudder and a backup in the same unit. In what was to be a belt-and-suspenders plan, Boeing told the FAA that in the unlikely case of a loss of both primary and secondary rudder control, pilots could still control the movement of the plane with the use of panels on the wings called ailerons.

But two aspects of the device in operation went unappreciated when the FAA certified the plane in 1967. First, the two cylinders in the PCU were "hand-fitted and mated for life," according to Haueter. Because the space between the two cylinders was hair-thin, just enough to allow one valve to move within the other, particles trapped between the cylinders could cause a jam that could then make the rudder go in a reverse direction, as when a steering wheel is turned to the right and the car goes left. It wasn't often, and it wouldn't happen in all the units. "Some would never reverse," Haueter said when we talked about the case years later. "Some were extraordinarily sensitive to reversal because of the way they were made."

When the rare event did occur, the second unforeseen problem could reveal itself. At slower speeds the ailerons would not have sufficient power to offset the force of the rudder. This was discovered only in 1994, when another Boeing 737 crashed, also under puzzling circumstances.

On September 8, 1994, USAir* Flight 427 plummeted from six thousand feet during what was up until that point a normal flight from Chicago to Pittsburgh. The plane was crossing the

---

* USAir was renamed US Airways in 1997. In 2012 it merged with American Airlines.

wake turbulence of a Delta Boeing 727. Suddenly it rolled to the left, surprising both pilots. Then the nose of the plane went down. In his book, *The Mystery of Flight 427: Inside a Crash Investigation*, author Bill Adair gives a horrifying description of the sensation experienced by the 132 people on board. "It would have felt as if they had reached the top of a roller coaster and were starting the first, huge drop."

The pilots pulled back on the yoke, but the left wing remained pointed down, and the plane began spiraling to the left, picking up speed as it fell. From the initial upset to the plane's crash into the woods just outside Pittsburgh, the event took twenty-eight seconds.

It was a difficult investigation. Flight data recorders at the time contained just a few details, but no data about the position of the rudder or ailerons. Tests of the components that control the plane's lateral movement did not reveal any problem that could have caused the plane to act as it did.

Throughout the investigation, however, Boeing insisted there was nothing wrong with the plane or the rudder. "The rudder was doing what it was asked to be doing," chief 737 engineer Jean McGrew told investigators in January 1995. McGrew was saying the pilots had mishandled the controls, exacerbating the problem by slamming on the rudder pedal and pulling back on the yoke, effectively stalling the plane.

Boeing clung to that position for two years. Then, in October 1996, Ed Kikta, a young Boeing engineer, was reexamining data from an earlier test and discovered that a fail-safe mechanism in the rudder control unit did not work as it should. Kikta was simulating a jam when hydraulic fluid began flowing in the wrong direction, which would then cause the rudder to swing in a direction opposite to what the pilot expected.

After a few hastily arranged follow-up tests, Boeing notified the FAA, agreeing to redesign faulty parts and supply them to 737 operators. Boeing started a campaign to help pilots understand how to react to in-flight upsets. John Cox, a 737 captain for USAir and a member of the accident investigation team, was enthusiastic about the pilot training. But he disagreed with the notion that Boeing would not abandon: the pilots' actions caused the accident.

Cox listened to the cockpit voice recorder repeatedly. He told me each time he heard it it was clear the pilots "had no idea what was going on. They never understood what the aircraft was doing or why it was rolling uncontrollably, and they could not stop it."

During an interview at the time with John Purvis, who was a Boeing air safety investigator, he took a similar tack, spinning the rudder fix not as a remedy to a problem with the plane but as an additional precaution. He told me, "We're making a safe plane safer." I was to hear the same thing two more times over the next fifteen years as Boeing recognized, both times under duress, that its designs could sometimes harbor hazards.

"Go to the engineering folks. They'll say, 'It couldn't happen, it's perfect. I understand that," Haueter told me. It's like what happens when the police show up and say your straight-A student robbed a convenience store. You're not going to believe it. Engineers are so embedded into the system, they know the design so well, they can't see its flaws."

I was working as a correspondent for CNN in 1996 when I got a call in the middle of the night telling me about the crash of TWA Flight 800. It was a Boeing 747 with two hundred thirty people on board that exploded thirteen minutes after taking off from New York on its way to Paris. There had been no distress call, just normal cockpit conversation—and then, wham, the plane broke into three large sections, leaving a long trail of debris in the Atlantic Ocean. Where each of these sections landed would help investigators determine the timeline of events, but not what triggered the blast.

Federal law enforcement agents were concerned about an act of terrorism, but the tin kickers with the safety board concluded the blast had been triggered by a flaw in the design.

The 747 jumbo jet and many other models produced by Boeing feature a large fuel tank in the space between the wings—the structural center of the plane. This tank was designed to double as a heat sink for the air-handling equipment located below. But it worked that way only when there was fuel in the tank to absorb the heat. When the tank was empty, there was nothing but fumes, which would heat up as if the tank were a giant saucepan sitting on top of the stove. It could get hot enough to ignite.

In my book *Deadly Departure*, I write that the stunning news to emerge from the four-year probe into the disaster was that this flammability problem was well known. Engineers at Boeing, some of the airline's customers, and federal air safety regulators had been discussing it for thirty-five years, because of a number of similar events beginning in 1963. In those cases, Boeing concentrated on locating the specific ignition source rather than the greater hazard of operating with a fuel tank in an explosive state.

Studies carried out as part of the Flight 800 accident probe showed that fuel tanks with heat-generating devices below them can be like ticking time bombs as much as one-third of the plane's operating time.

In the 1960s and '70s, safety officials asked for devices to be installed in fuel tanks to preclude the possibility of explosion by eliminating the oxygen, a necessary component of fire. During the development of the 747, Boeing even tested some systems specifically designed to do this. The manufacturer ultimately dismissed the idea, however, citing concerns about the additional weight.

The recommendations for protecting the fuel tanks emerged repeatedly over the decades, but the FAA accepted Boeing's position that if the triggers for the explosion could be identified and fixed, the design would be safe enough. What the TWA 800 disaster showed was that there would always be unknown triggers. We would have to call them the "known unknowns."

In 2006 the U.S. Department of Transportation issued a new rule: all new airplane designs had to include a system to protect the tank from explosion. Boeing's newest airplane, the 787 Dreamliner, incorporates a fuel tank explosion-prevention system that is a direct result of the TWA crash investigation.

So it is ironic that after just fourteen months in service, the 787, which had already secured its place in the pantheon of revolutionary aircraft, was gracelessly sidelined for four months in 2013 because of a risk of fire and explosion from the plane's lithium-ion batteries.

The designers of the Dreamliner and the Comet shared an overconfidence in their creations. "The Comet embraced new technologies before they were fully understood," Graham Simons, the author of *Comet*, told me. "With the Dreamliner, Boeing pushed the same limits. They seem to have forgotten when you push the envelope, you open a greater area of risk."

Fever Dream

Among Boeing's worldwide customer base, there are few as loyal as Japanese carriers Japan Airlines and All Nippon Airways. ANA effectively launched the 787 Dreamliner by placing the very first order for fifty of them in 2004. Ten years later it remained the largest 787 operator.

There was much national pride in the 787 because of the number of Japanese companies making parts for it. Fuji Heavy Industries, Mitsubishi Heavy Industries, and Kawasaki Heavy Industries all turned out pieces that were shipped to Boeing's American assembly plants.

While these giants of Japanese industry made the big parts, in Japan's historic former capital of Kyoto, battery manufacturer GS Yuasa was churning out a much smaller and more obscure component that was on the cutting edge of transportation power systems. Lithium-ion batteries the size of a bread box and the weight of a dorm refrigerator were to ignite the biggest issue in aircraft design since the Comet.

The first time I saw a Dreamliner outside the factory, it was on its six-month world tour. The aircraft arrived at Bole International Airport in Addis Ababa in December 2011. Ethiopian Airlines captain Desta Zeru, dapper in a forest green uniform, was at the controls for the first flight of the new airliner into the African continent. Ethiopian was a launch customer, with ten 787s on order. Ethiopian was also a new member of the Star Alliance, the world's largest network of airlines, so the airline hosted a three-day extravaganza to celebrate. Boeing created a large meeting space on board the airplane and invited reporters in for a press conference and look-see.

Of course, everybody was delighted to have the plane on display. Three years late to customers, the plane was sometimes called the seven-*late*-seven, and Boeing had had it up to here with criticism of its inability to set a delivery date and stick to it.

In late October 2011, however, when the first Dreamliner began revenue flights for All Nippon Airways, the game-changing airplane was changing headlines, too. "The engines purred rather than roared." Travelers were "agog." The engines "sipped fuel," and the passenger cabin "glowed," and that's just one review* from among thousands written in a similar vein once the plane actually started flying. President Barack Obama called the Dreamliner "the perfect example of American ingenuity."

The Japanese are enthusiastic air travelers, and they embraced the Boeing 787, too. Kenichi Kawamura is a policy adviser to his father, a Japanese political official in Tokyo. His job required him to commute by air each week between his home in Yamaguchi and his father's office. A self-described aviation enthusiast, Kawamura knew he was traveling on a special airliner on January 16, 2013.

---

* John Boudreau, "In Praise of the 787's Emotional Experience," *San Jose Mercury News*, September 25, 2012.

Eighteen minutes into the ninety-minute flight on ANA's Flight 692 to Haneda Airport, the Boeing 787 Dreamliner nosed down precipitously. Kawamura grabbed his drink moments before it would have tipped off his tray table. He had never experienced such a rapid shift. "It was a sudden fall and very steep," he told me. Looking at the flight profile, an experienced airline pilot described it as being like riding in the backseat of a car racing down a highway and then suddenly having the driver hit the brakes.

"Then there was a smell like plastic burning," Kawamura said. The flight attendants were walking up and down the aisle purposefully, collecting the cups and bowls in which they'd recently served miso soup. As an attendant approached his row, Kawamura started to ask what was going on, but then stopped as a voice came on the PA. "This is the captain," Kawamura remembers hearing. In fact, it was the forty-six-year-old first officer, one of the airline's very first pilots to be certified to fly the Dreamliner.

"We have smoke, we smell smoke," Kawamura remembers being told. "We must make an emergency landing." That much was already clear. Kawamura was heartened to hear the pilot say that instruments in the cockpit indicated that there was no problem.

In truth, the pilots were worried that the flight instruments were giving bad information. As soon as the first officer was finished with the announcement, he told air traffic control that there was "thin smoke, possibly by an electric fire." The main battery had failed. The pilots wanted to land as "soon as practicable," he said.

The drama for the flight crew began sixteen minutes into the flight, as they were taking the plane through thirty-two thousand feet on the way to leveling off at forty-one thousand feet. Voltage dropped in the main battery used to provide emergency power in case of the loss of other systems. It was not a subtle decline, either, falling from thirty-one volts to eleven volts in ten seconds. The pilots were unaware of this, though. Their first alert was an advisory that emergency floor and exit lights had come on in the cabin.

The first officer had only moments to wonder what might have triggered the lights, because everything seemed to be working fine. Then he smelled something burning.

Pilots adopt an air of bravado about their work. But if there is a kryptonite for aviators, it is fire. That's a "no-shit problem," Capt. James Blaszczak, the retired Dreamliner captain, told me in reference to how the ANA crew would have reacted.

Nine days earlier, a Japan Airlines Boeing 787 (JA 829J) with just 169 flight hours and 22 landings was parked on the ramp at Boston's Logan International Airport when a maintenance worker called the airline's station manager, Ayumu Skip Miyoshi, reporting "smoke inside the cabin."

At first Miyoshi was confused. "So you mean one of the passengers smoked in the lavatory?" he asked. "Smoke inside the aircraft cabin" was the reply. Miyoshi hurried out to the tarmac, and as he approached, another maintenance worker was running toward the plane with a fire extinguisher in his hand. Smoke was billowing out of the electronics bay. When firefighters arrived, they said the battery was hissing, sputtering, and popping as flames leapt from the connectors on the blue battery case. It took an hour and forty minutes to get the smoke under control. The contents of the battery box continued to burn until all the fuel feeding the fire was exhausted.

Every operator of the Dreamliner heard about the Boston

event within minutes. I was interviewing former American Airlines chief Robert Crandall at his Florida home when his wife interrupted us with the news. CNN was carrying the story live. Forty-seven airlines around the world had already ordered the airplane. If Crandall's reaction was typical of other aviation executives, they were all glued to their TV sets.

A number of entry-into-service problems, including fuel pump leaks and engine fan shaft fractures, had been reported but were considered ordinary shake-out stuff that operators come to expect with new designs. The battery fire, however, was no minor issue.

Fire, smoke, and airplanes don't mix—ever. Or as Captain Blaszczak told me, "If there is any indication of smoke or fire, the definition of eternity is from now until we get this airplane on the ground."

Fire is not a common occurrence in air disasters. About 16 percent of commercial aircraft accidents involve fires; less frequently are they the cause. That is because a lot of effort has gone into minimizing the risk, which is a difficult task, considering that combustion is what makes the engines run. The attention paid to fire prevention is so great because an in-flight fire can quickly turn to catastrophe.

On a Swissair flight from New York to Geneva in 1998, pilots mistook smoke in the cockpit as a minor air-conditioning issue. Trying to identify the source delayed by four minutes how they dealt with what turned out to be an electrical fire in the space above the cockpit ceiling. Arcing, a high-temperature electrical discharge across a gap in wiring, had ignited highly flammable insulation material, and the fire spread quickly. Just twenty-one minutes passed between the first smell of smoke and the plane slamming into the Atlantic Ocean off the coast of Nova Scotia, killing all 229 aboard.

In its report on the accident,* the Transportation Safety Board of Canada concluded that pilots dealing with fire have between five and thirty-five minutes to get the plane on the ground. That's it.

Swissair 111 "was a seminal event in aviation history," said Jim Shaw, an airline captain who participated in the accident probe on behalf of the Air Line Pilots Association. There were a lot of lessons learned. One of the most important to Shaw was the error in the FAA's decision to allow McDonnell Douglas to continue to use metallized Mylar (polyethylene terephthalate, if you really want to know) to insulate the walls of its airliners, even after other safety authorities determined it was flammable.

It was a curious decision. In the decades leading to the accident, airplane manufacturers had been ordered to replace all sorts of interior fabrics and materials with those that were fire-retarding and flame-resistant. Seats, dividing walls, carpets, curtains—all these had to be made of substances that were slow to ignite and self-extinguishing, meaning that if a fire were to start, it could not spread more than a few inches. The rules went into effect on all planes built after 1990 and included the MD-11 that flew as Swissair Flight 111.

So, naturally, you are wondering if the metallized Mylar was flammable, why was it on the Swissair airplane in the first

---

* "Aviation Investigation Report A98H0003," Transportation Safety Board of Canada, http://www.tsb.gc.ca/eng/rapports-reports/aviation/1998/a98h0003/a98h0003.asp.

place. The answer is that it wasn't prohibited because it was used away from fire zones. Absent an ignition source, the flammability of the insulation material didn't matter, or so the thinking went at the time.

It did matter, though, because power cables, light fixtures, battery packs, and electrical wires all ran throughout the insulated area. A spark from any one of them could trigger a fire, which is what happened with Swissair 111.

After the crash, safety regulators said metallized Mylar had to be removed from twelve hundred airplanes. The effort took years, and by the time the last of it was being pulled out, Boeing was trying to convince the FAA that another highly flammable material, lithium-ion cobalt-oxide batteries, should be allowed to power the new plane the company was designing, the Boeing 787 Dreamliner.

Powerhouse

In the post-9/11 world, the twin-engine, mid-capacity, long-haul airliner on Boeing's drawing board was the most eagerly anticipated airplane since the Boeing 747, which had redefined travel in the 1970s by opening up the skies to everyone. Forty years later, the 787 was going to allow airlines to fly between far-flung secondary cities without having to rely on masses of passengers to fill the plane. Sure, any airline could sell four hundred fifty seats on a London-to-New York flight, but the markets between Houston and Lagos, Auckland and San Francisco, Toronto and Tel Aviv, would be thinner. A smaller and more fuel-efficient airplane that could fly eight thousand miles, nearly one-third of the way around the globe, would be a showstopper.

Two things made the Dreamliner different from any other airliner at the time. The first was that Boeing had eliminated vast amounts of aluminum in the plane's structure, just as de Havilland had done decades earlier. Where de Havilland simply used thinner metal (and wound up making it thicker in later models), Boeing replaced aluminum with stronger but lighter-weight carbon fiber.

The second differentiation was all twenty-first century: the plane would have an unparalleled reliance on electricity. Six generators would convert energy from the engines to electrical power, creating five times as much as any other airliner, enough to power four hundred homes, the Boeing press material claimed. The 787 would be a virtual power plant in the sky, supplying its own needs, and they were mighty. Boeing replaced mechanical flight controls and their heavy stainless-steel cables with electromechanical controls that activated cabin pressurization, brakes, spoilers, the stabilizer, and wing ice protection.

A critical part of this new energy production, storage, and distribution plan was the use of two powerful lithium-ion ships batteries, which would provide start-up power and supply some emergency electronics, lighting, and independent power for the black boxes.

Not all lithium-ion batteries are alike. There are manganese, iron phosphate, titanate, sulfur, iodine, and nickel. Each element or compound offered some benefit and some drawback. The names refer to the combination of materials used to move ions from one chemically coated thin strip of metal, through a thin permeable sheet, to another chemically coated strip, all of which is wound into a jelly roll shape and placed in a metal can, then sealed and called a cell. The motion of the ions generates electricity. The electricity is collected and stored until it is called on to deliver power. And now you know how batteries work.

Of all the lithium-ion battery recipes, the one called cobalt oxide had the best energy-to-mass ratio, meaning it gave the most energy for the least weight and size. It had other features that made it desirable: It charged quickly and held a charge longer than others. It was slightly more expensive than lead-acid batteries, but because it was already a widely produced item, the cost was below that of any of the other chemistries.

Jeff Dahn, a professor of physics and atmospheric science at Dalhousie University in Halifax, pointed out, however, that for what Boeing was trying to achieve, cobalt oxide was "not well suited" because it "has inferior safety properties compared to other alternatives." He was referring to what has been reported to be the largest industrial recall ever.

In 1991, Sony Energy Devices of Japan held the patent for one of the formulas used to produce cobalt-oxide lithium-ion. About thirteen years later, it was a multibillion-dollar product, powering all kinds of what are called 3 C devices: those used in computer, communication, and consumer electronics.

The batteries had a tendency to heat up on their own, progressing to fire and explosion, according to a paper produced for the International Association for Fire Safety Science in 2005. Some of the spontaneous combustion events had been filmed and uploaded to YouTube, where anybody could view alarming videos of cell phones and laptops sputtering and emitting tongues of flame in airport boarding areas and at office meetings.

At the time Boeing selected lithium-ion, in 2006, the U.S. Consumer Product Safety Commission was issuing recalls for the batteries powering the devices sold by Lenovo, Dell, Toshiba, Apple, and others. About four million battery packs sold by Dell alone were ordered off the market in August of that year. "Consumers should stop using these recalled batteries immediately and contact Dell to receive a replacement battery," one commission recall notice read.

This was all big news, but at the same time at the FAA, Boeing was pushing a plan to use lithium-ion batteries on the Dreamliner. It was progressing through the bureaucracy with some caution, but little public attention. The FAA told Boeing that if it wanted to use the technology, it would have to meet the terms of a special condition, because nothing in the existing regulations governing airliner design addressed this "novel technology."

There was limited experience, according to the FAA, and what experience there was wasn't good. When the FAA asked the public to chime in, just one organization did, the union representing the pilots who would ultimately fly the airplane.

"We got involved because we've always had issues," said Keith Hagy, the ALPA safety director, explaining why the union wrote several letters to the FAA during the process. Hagy wasn't so confident Boeing could achieve safety on the 787 while using batteries with such a troubled past. "Once they start burning, they never go out," he said.

ALPA was not standing on a deep well of research. So sparse was the material publicly available that it relied on just one document, the FAA's "September 2006, Flammability Assessment of Bulk-Packed, Rechargeable Lithium-Ion Cells in Transport Category Aircraft," and that wasn't even focused on powering an airplane with lithium-ion; it was about how airlines should pack the batteries if they were being sent as cargo. What struck Hagy is that even Boeing didn't know much about the batteries.

Whether the information wasn't there or the effort to get it was ineffective, people in the battery industry dispute the notion that lithium-ion batteries were a new frontier. Maybe they weren't being used on airplanes, but they'd been under review by NASA since 2000. Boeing's space division generates ten billion dollars a year, roughly 12 percent of the company's total revenues on average, but according to one scientist with knowledge of the battery development, no one from the company's space side was ever asked to provide information or expertise to the commercial airplane division or even to review the Dreamliner battery design until after the events in January 2013.

Nor does it seem that the experience of automakers was sought even though they had been working on electric cars for nearly a decade. The Tesla Roadster, a high-end electric car, was being developed in 2006 using the same cobalt-oxide formulation as Boeing, but the two companies were not sharing information.

Through the summer and into the fall of 2006, the Consumer Product Safety Commission's recalls continued. One computer company after another was telling customers to stop using the batteries that came with their laptops and to request a free replacement. And just as doggedly, Boeing kept working on the design of the lithium-ion battery system it would put on the Dreamliner. The day before Halloween, another round of recalls of nearly a hundred thousand batteries was announced. One week later, on November 7, 2006, the news became personal for Boeing.

A fifty-pound prototype, so expensive to produce and so chock-full of power it was called the "Ferrari of batteries," had been delivered to the Arizona headquarters of Securaplane Technologies from the Japanese maker GS Yuasa. It wasn't the final product, but it was what Securaplane was going to use to test the charger, its contribution to the power generation system on the Dreamliner.

Michael Leon, a technician at Securaplane, was one of those

assigned to work on the test, but the battery made him nervous. Earlier, there was a short circuit between the terminals. Workers immediately removed the current, but not quickly enough to prevent a second short circuit. Leon didn't want to use the battery again, but executives at GS Yuasa were untroubled. Analyzing the data sent by Securaplane after the first short circuits, the Japanese said that, with proper handling, the battery would be fine for continued use.

Then, on November 7, the battery ignited and exploded. Leon told the *Arizona Daily Star* that flames were leaping ten feet in the air. "The magnitude of that energy is indescribable," he said. No one was hurt, but the company's administration building was destroyed. For everyone involved, it was a multimillion-dollar lesson in unknown dangers. Boeing reevaluated its selection of lithium-ion and considered swapping it for lithium-manganese, but didn't.

Everyone went back to work: GS Yuasa on the battery, Securaplane on the charger, and the French company Thales, which had been hired by Boeing to oversee it all.

Primarily, everyone working on the battery design was concerned that in the process of charging the battery, too much power had been pushed into a cell, causing it to heat up and ignite. So they added a contactor that would disconnect the battery from the power supply. It would take the battery out of commission, bricking it for the rest of the flight, but that was considered the lesser of two evils.

In April 2007, the FAA published a two-page special rule in the Federal Register that would give Boeing the go-ahead to use lithium-ion batteries on its new airliner, provided it met certain conditions. The FAA addressed the battery's persnickety nature using aviation's four letter F-word, *fire*, and some other bad words, such as *flammability*, *explosion*, and *toxic gases*. Boeing would have to make sure cells didn't heat up uncontrollably, cause the failure of adjacent cells, catch fire, explode, or emit toxic gases. The plane maker was still a long way away from being able to claim it accomplished this, and in fact, it never could. Then, in July 2009, a new hurdle emerged.

Engineers at a Hamilton Sundstrand lab in Rockford, Illinois, were plugging together 787 hardware to see how it all worked as a system. The answer was "not well." One of the cells heated up uncontrollably, spewing electrolyte and causing the entire battery to fail. Boeing had a second round of second thoughts. Lithium-manganese and nickel-cadmium battery alternatives were briefly put back on the table. Yet, really, "ni-cad" offered only a little more than a tenth of the power for start-up as lithium-ion, and the company still didn't like manganese, so once again Boeing stuck with its original choice.

"If they understood the risks, they never would have done it," said Lewis Larsen, a Chicago entrepreneur and theoretical physicist whose work requires him to know about this chemistry. In 2010 he sent a presentation to Boeing and all the automobile companies with similar plans to use cobalt-oxide lithium-ion batteries for power. Through his company, Lattice Energy LLC, Larsen has been tinkering with lithium-ion because of the very characteristic that makes the batteries so inherently unsafe: the microscopically small, naturally occurring dendrites that grow inside cells over time, creating a pathway for internal electrical shorts called field failures. These miniature flash fireballs generate temperatures between five thousand and ten thousand degrees Fahrenheit. They are called LENRs, for "low-energy nuclear reactions." Larsen studies LENRs because he thinks they can be used as a source of green energy.

As a feature of a battery that will be used on an airplane, however, the idea of flash fireballs ought to set off alarm bells. Writing for the *Encyclopedia of Sustainability Science and Technology* in 2012, Brian Barnett said that field failures could create "violent flaming and extremely high temperatures" as well as explosive combustion. "Most safety tests carried out in the laboratory or factory do not replicate the conditions by which safety incidents actually occur," Barnett wrote.

This was the nature of the message Larsen sent off to Boeing. He told me, "We thought it was a moral issue to make some public statements about what Lattice knew technically at that time, so we did." Larsen said he heard back from a contact at the company, who told him that ten battery experts had assured Boeing that Larsen "was full of shit."

Thermal Fratricide

In the fall of 2011, the first Dreamliner was delivered to All Nippon Airlines, complete with the two ships batteries that had been the subject of so much tinkering. The following spring, the airline reported that not only was it pleased with the airplane, but its customers were also. Nine out of ten passengers said the 787 met or exceeded expectations, ANA reported. Who pays close attention to the model of airplane on which they fly? The Japanese. Eighty-eight percent of Japanese travelers surveyed were familiar with the Dreamliner when they boarded their flight.

Koichi Hirata, fifty-seven, was on his way to attend the InterNepcon electronics show in Tokyo on January 16, 2013, and, like those other happy ANA customers, he was looking forward to flying in the 787. He had a business-class window seat on the right side of the airplane, and he was listening to Rakugo, a traditional form of Japanese comedic storytelling on the in-flight entertainment system. He noticed right away when the plane turned and started to descend.

"The passengers didn't get into a panic, and cabin attendants seemed to be making emergency landing preparations with great efficiency," he recalled later. Once the plane was on the ground in Takamatsu and stopped in a cleared area of the taxiway, Hirata saw smoke being drawn into the engine on the right side, behind where he was seated. He didn't have much time to think about it because the flight attendants were telling the passengers to leave their things, take the emergency slides, and get away from the plane. Hirata said that between the smoke and the rapid rush to the exits, he wondered if the incident was "something far more serious." As far as he and his fellow passengers were concerned, the answer was no. Yet for Boeing, ANA, and the four dozen other airlines that had invested in this airplane, the answer was an unequivocal yes.

The JAL battery fire in Boston nine days earlier had been alarming, but until this ANA episode in Japan, the reaction of aviation officials had been to suggest that the battery meltdown was a one-off. The then-administrator of the FAA, Michael Huerta, had even gone so far as to hold a reassuring news conference with Boeing CEO Ray Conner by his side, at which he said that "nothing we have seen leads us to believe the airplane is not safe." That was January 11, five days before the emergency landing at Takamatsu. The second event resulted in the decision to ground the fleet worldwide.

In just over a week, the world's newest airliner experienced two incidents that the safety board's then-chairman characterized as unprecedented and serious. "We do not expect to see fire events on aircraft," then-NTSB board member Deborah Hersman told reporters.

It had become clear to Hersman that the lithium-ion cells

in the Dreamliner battery box—the cells Boeing was supposed to coddle like a temperamental child, the cells the FAA warned had characteristics that "could affect their safety and reliability"—had just done the very thing they were not supposed to do under any circumstances: heat up uncontrollably. In Boston, which was the incident the NTSB would investigate, this had caused a very-high-temperature chain reaction that destroyed the entire eight-cell assembly.

"These events should not happen as far as design of the aircraft is concerned," Hersman told reporters. "There are multiple systems to protect against a battery event like this. Those systems did not work as intended. We need to understand why."

This was not as easy as it sounded. Every air accident may be different, but there are common investigative themes: mechanical, operational, organizational. The mystery on the Boeing 787 wasn't about the physics of flight or the human operation of the plane. This was a puzzle about complex electrochemistry. When Hersman's investigators in America and their counterparts in Japan set about to educate themselves on the topic, they quickly found that many of the experts had already been hired by Boeing. Though it had eschewed outside advice during the development phase, the company was now assembling advisory committees and review boards and scooping up spare batteries for its own testing.

Dana Schulze, deputy director of aviation safety at the NTSB, said getting up to speed for her investigation was frustrating. Still, she showed remarkable understanding of Boeing's point of view.

"Our job is safety, but it's hard to argue that the priority was getting this plane back in the air," she told me. "At the same time, we had an investigation to conduct and wanted to be sure that the safety issues were addressed."

Judy Jeevarajan was one of those experts who found herself on the receiving end of a call from Boeing. From 2011 to 2015, she was battery group lead for safety and advanced technology at NASA. Part of her job was to study and publish articles on how to use lithium-ion batteries safely in a manned space environment. Because the batteries are so much larger and more powerful than those used on personal devices, when they go bad, they have the potential to trigger catastrophe. Jeevarajan's job was to manage those risks on the International Space Station by critically assessing the battery at three levels: the individual cell, the cells as a unit (the battery), and the electrical system into which they were integrated.

When the Dreamliner was grounded, Jeevarajan was recruited by Boeing to give an honest opinion about the battery design and function. On the day the experts got together in Seattle, the damaged battery from the JAL plane in Boston was wheeled into the room on a cart. Everyone crowded around for a look. Jeevarajan was surprised at what she saw. The cells were missing any bracketing substantial enough to keep them firmly in place, protected from vibration, and separated from one another. This observation was shared by Kazunori Ozawa, a battery engineer who was involved with the development of lithium-ion batteries many years before at Sony, but who was not part of the review. He had seen photos of the damaged battery.

"By looking at the inside of the pack after the accident, it can be easily understood that those cells were not clamped well," Ozawa told me. He was concerned that the cells would vibrate in an aviation environment, causing the jelly roll–like windings to move around inside the case, perhaps even triggering short circuits if the terminals touched.

Also, Jeevarajan noted that hot spots could develop inside each cell and within the foil/chemical roll if "used outside of manufacturer's specifications." Substantial heat could melt the film separating the cathode from the anode, a big no-no because it can trigger a short circuit and immediate thermal runaway.

Professor Dahn also thought that was a problem. Referring to tests by the NTSB during the investigation, he noted that "when the cells were discharged at their max-rated power, the terminals were hot," he said. "They became one hundred eighty Celsius [three hundred sixty degrees Fahrenheit] above the melting point of the separator. That's bad." As far as he was concerned, part of the problem was the size of the cells. At two inches each, the individual cell pack was just too thick. One failure might propagate and "set off the neighbors," as he put it.

GS Yuasa designed the Dreamliner battery based on one it produced for all-terrain vehicles. When I met with Dana Schulze at the NTSB's offices in Washington, I asked her if it had been a leap to suggest an ATV battery and a battery on an airplane were the same. She said GS Yuasa's experience could be applicable as long as the differences between the two were considered, something Schulze, a mechanical engineer, called a similarity analysis. In the NTSB's review of the battery design, "We couldn't find evidence that they had done a lot of that."

Locked in competition with Boeing with its same-size and same-range Airbus A350, it might sound like bragging when the French airplane maker Airbus points out the various differences between the 787 battery and the lithium-ion battery now powering the A350. But, actually, I called Airbus, not the other way around.

Airbus accounted for the possibility of an internal short circuit in one cell, according to Marc Cambet, a systems component architect for the A350. The design included precautions to prevent one failing cell from spreading to another. "On our side, the cells are insulated one from each other and insulated in terms of thermal insulation and in terms of vibration and in terms of separation between the cells," Cambet said.

Jeevarajan's review of the Boeing battery covered many issues, including propagation, insulation, and vibration. She had one overriding problem, though, and that was Boeing's safety philosophy when it came to the batteries. The company worried too much about overcharging them and ignored the potential for and response to internal and external short circuits. These could result from something such as the melting of the separator; the build-up of dendrites; or field failures, internal pressure, and deformations in the cell case.

It was not that Boeing failed to consider them, Jeevarajan said. The company had, but it had also dismissed the hazards. When it tested for external shorts, "every single cell" vented and emitted smoke. When she asked the company, "Why didn't you take that into consideration?" the answer was, "There was no fire. We didn't think it was a big deal."

"It blew my mind that they had every single cell vent with smoke," which was expressly prohibited under the FAA's conditions, she reminded me, "and they just ignored it."

"They had good protection for overcharge, but none for short circuits," Jeevarajan concluded, and overcharging wasn't a factor on either of the Japanese airliners. In the pages of notes she gave to Boeing, Jeevarajan spelled out what she saw as the design's many shortcomings, including the one that the Japanese investigators would seize on much later as the likely culprit for what happened on the ANA Dreamliner.

"If you want to charge at cold temperatures, you've got to reduce the charge current," she explained, or risk generating too much heat. It's harder for the ions to move when it's cold, so charging generates more heat in the battery.

On the A350, the battery temperature is taken into account, Cambet explained. "The charge is slowed when the temperature is too low," he said, because "if you try to charge full power at low temperatures you can build some risk of internal short circuit."

Moderating the rate was not possible on the Dreamliner battery system, and the fact that the ANA and JAL events happened in January was an interesting piece of the puzzle. Around this time, one of the government investigators (who asked not to be identified by name) summed it up, saying, "The cause was not found, but so many potential causes were found that [it] was pretty surprising."

In Tokyo, as in Washington, DC, the air safety folks were trying to solve a mystery while taking a crash course in electrochemistry. The Japan Aerospace eXploration Agency (JAXA) was advising, as were NASA and the U.S. Naval Sea Systems Command, which, like the space programs, is intrigued by the potential of these batteries, yet cautious about the hazards.

"It was quite difficult to take enough time to learn the new technology," said Masaki Kametani, fifty-three, one of the seven investigators assigned to the case in Japan. Their little department was the subject of international attention. While the Boston battery fire had taken place on the ground in a nearly empty plane, the ANA flight had been in the air with passengers. "What if" was on everybody's mind.

The investigations' official reports ran to hundreds of pages. Japan's took twenty months to complete, and the American version nearly two years. While the flaws in the design of the battery were many, neither agency could say specifically what initiated the incidents they'd investigated. The JTSB concentrated on the cold weather charging phenomenon. The NTSB thought a manufacturing defect in the cell case might have created a hot spot leading to a short circuit. In both cases the entire battery fell to what the industry and physicist Lewis Larsen graphically call "thermal fratricide."

In its report, the NTSB went beyond what had happened on the plane in Boston to review how Boeing had convinced the FAA that the battery would meet the special condition. There were dozens of tests with a mind-numbing collection of titles and findings. Yet later, when I interviewed the NTSB's Schulze, she explained it to me quite simply. It wasn't the number of tests that was relevant; it was the assumptions made by Boeing as to what actually needed to be tested and how closely those checks reflected how the battery would operate in an actual airplane. "The test in and of itself really didn't provide enough information to assume that an internal short circuit wasn't going to result in propagation to the other cells," Schulze said.

The parallel to the Comet was unmistakable. The de Havilland engineers sought reassurance by referring to demonstrations of the cabin withstanding twice the amount of pressure the plane would experience. Yet during testing, the structure had been given supplemental support. It was therefore not representative of the plane in flight.

By the time the investigators published their findings, the

Dreamliners had been back in the air for nearly two years, released to fly again after a number of changes that recognized the battery's newly exposed hazards, if not its suitability. Boeing also acted on the suggestions of experts such as Jeevarajan and put more insulation between the cells, placing them in a stronger frame and isolating them electrically with Kapton tape. These were precautionary measures. They didn't change the battery's volatile nature, so Boeing opted to cage the beast.

Each battery went into a stainless steel housing with a titanium vent tube. The box would contain and smother a fire and protect against heat. The vent would release outside the airplane smoke or fumes generated by a failure. The system was an insurance policy against more unknowns, known or otherwise.

Boeing opted not to talk to me for this book, rejecting repeated requests made in person and via e-mail. I didn't like it, but I certainly understood. The company might be concerned that the plane it spent a decade developing will be permanently associated with one particular design failure. Had someone at Boeing given an interview, I'm sure that person would have reminded me that the 787 is carrying out the mission for which it was created. Airline customers love it, and passengers do, too. One thousand had been sold by 2015.

What Boeing probably believes in its corporate heart, is akin to Wilbur Wright's thoughts about "mounting a machine and becoming acquainted with its tricks." Progress involves risk. Innovation will always have unanticipated consequences. The challenge for plane makers past and present has not been finding the guts to gamble, but balancing audacity with prudence before the plane moves from design to product.

The price of getting it wrong can be too high, as Rolls-Royce and Lockheed learned in the early 1970s during the development of the L-1011 jumbo jet.

Ask any airline pilot who has ever flown it to tell you about the Lockheed L-1011 and be prepared for a long soliloquy. “Best seat in the sky,” one captain told me. Passengers enjoyed its unprecedented space and the quiet that had L-1011 customer Eastern Airlines calling the plane the “Whisperliner.” Yes, everybody loved the three-engine Tristar, but unlike its competitors, the Boeing 747 and the McDonnell Douglas DC-10, the L-1011 is not seen flying anymore. Only two hundred fifty were made, and aviation analyst Richard Aboulafia, who writes the widely read industry newsletter *Teal Monthly*, cites an engine design gone terribly wrong with contributing to the ultimate failure of the L-1011.

We hear about the use of composite material in aviation all the time these days, as it replaces heavier metal in airplane structures and components in planes such as the Dreamliner, the Airbus A350, and the A380. Back in the sixties, however, when Rolls-Royce was designing a more powerful, lighter-weight engine to power wide-body aircraft, its plan to use woven layers of compressed and hardened glass fibers to replace the metal on engine fan blades was new. Using a composite called Hyfil for the blades would save three hundred pounds and make the engine 2 percent more fuel efficient, but it didn't work as planned.

In a 2012 *Royal Aeronautical Society* magazine story called “Blades of Glory,” writer Tim Robinson called it “the blade that almost broke the company.” While the engine was still in development, Rolls-Royce and Lockheed discovered that the Hyfil composite was prone to delaminate. The layers would separate, just like what happens when a laminate board is left out in the rain. The fan blades would also not withstand the assault of frozen poultry. That's not as bizarre as it sounds; to ensure that an engine will survive ingesting birds without coming apart, manufacturers shoot hard-as-ice frozen chickens into the core of a spinning jet engine. In this case, it was a battle the roasters won.

While Boeing and McDonnell Douglas churned out their wide-bodies, sending them to airlines so they could start flying to far-flung places, Lockheed's L-1011 was stuck on the ground waiting for Rolls-Royce to refit the engine with titanium blades. Unable to pay its bills, Rolls went into government receivership. Lockheed fared only slightly better; it required government loans to stay in business. It took years for Rolls-Royce to finally deliver the engine Lockheed had ordered for the L-1011.

Aboulafia says the bad bet on Hyfil was orders of magnitude worse for the companies involved than what Boeing faced with its lithium-ion batteries, even though the engines should have been an easier fix. "You have to distinguish between two types of design. On the Dreamliner the issue is fundamental, and on the L-1011 the problem was an accessory; an expensive accessory, yes, but it was a discrete system," he said. "Replace the engines, problem solved; it was not a fundamental design flaw."

The problem for Lockheed was that it was locked into the Rolls-Royce engines; it could not swap them out for the product of another engine maker. Lockheed had no choice but to wait out the delay. Once the composite blade was replaced with titanium, the RB221 became one of Rolls-Royce's best-selling engines. Updated versions still power the Boeing 747 and Boeing 767.

The Dreamliner battery saga does not have such an unequivocal finish or even a dignified one. Aboulafia calls the stainless steel containment box "inelegant," the opposite of a Rube Goldberg invention, where a comically over-engineered response addresses a simple quandary. "It's a simple solution to a complex problem," he said. And it is a problem that just won't go away.

On January 14, 2014, nearly a year to the day after the ANA Dreamliner made its emergency landing at Takamatsu Airport, maintenance workers in the cockpit of a Japan Airlines Dreamliner preparing to depart Narita for Bangkok saw white smoke coming off the plane. Checking the battery, they discovered that one of the eight cells had vented, leaking fluid into the box. The problem did not spread to other cells.

Ten months later, a Qatar Airways 787 had to make an emergency landing when one cell in the airplane's battery vented. Boeing notified the NTSB and the FAA, but neither agency conducted an investigation.

"The airplane performed as certified for this failure," FAA spokeswoman Laura Brown told me. When I asked by what means the air safety authority would know how the airplane performed since it had not investigated, Brown said that Boeing had taken a look and shared its findings.

Not only is the FAA not investigating when cells vent, but it is not even keeping count of how often it happens. Since the box contains the smoke, fumes, and presumably fire, the American aviation regulator's position is that a battery malfunction is no longer their concern. "We don't consider something an 'event' if the containment box performs as designed," she said.

Several battery scientists, including some who do not want their names to be associated with their opinions because they work with Boeing, say it is lunacy to dismiss the seriousness of continuing cell failures.

John Goodenough, a physicist and professor at the University of Texas who is considered the inventor of lithium-ion batteries, points out that by the time electrolyte vents, there is fire within the cell. "If you are hot enough to start boiling and needing to vent, the electrolyte will have caught fire by then."

Since it first approved Boeing's use of lithium-ion energy storage on the 787, the FAA went from saying that fire in any situation was unacceptable to calling containing a fire no big deal.

"Why play with fire when you don't have to play with fire?" asked Dalhousie University's Jeff Dahn. That cells vented on four or maybe five batteries in the plane's first three years of service means the GS Yuasa–produced cells were failing at an "astronomically high rate," Dahn said.

Here's how he figures this. Each plane has sixteen cells, and by the end of 2014 there were fewer than three hundred Dreamliners in service. If the cell events noted here are the only ones, and we can't know, because no one aside from Boeing seems to be counting, and Boeing's not saying, then the failure rate in the cells is one in every few hundred. By contrast, Dahn said, "for the cells used in laptops and phones, the failure rate is one in more than twenty million. It's irresponsible to continue with such a product."

The FAA is not alone in its conviction that Boeing has solved its battery problem. In Japan, JTSB spokesman Mamoru Takahashi told my researcher Takeo Aizawa that the safety board was not surprised or concerned by subsequent cases of cell venting. This was similar to what the board's Koji Tsuji had told me. "We cannot create the situation where no short circuit happens, but we have reached the point where we can control the heating, even if venting occurs," he said. "We can consider it a minor problem."

"Boeing has made all the conceivable improvements for their 787s to ensure the resumption of normal flight," including more than eighty modifications that "are desirable, but may not be necessary," Tsuji told me. It was eerily evocative of what was said about the Comet before it was returned to service with the source of its problem still a mystery.

PART FOUR

Humanity

An accident sequence is like someone slipping down a knotted rope. The pilot's decision may be the last knot in the rope, but there are many other events which set up the accident sequence. Pilot error is increasingly seen as far too simplistic.

—MAURICE WILLIAMSON, NEW ZEALAND'S
MINISTER OF TRANSPORTATION, 1999

The Right Stuff

Sometimes in the summer I go kayaking with my friend Pete Frey, a pilot for a U.S. carrier whom you met earlier in this book. For a man in his late fifties, Pete's in great shape. Still, it's hard for men of a certain age to pull off the baggy shorts and windblown hair look. Truth is, Pete doesn't even try to look hip.

In his pilot's uniform, however, Pete is a babe magnet. It's not just the navy jacket with the gold braid; it's his confidence, his completely unselfconscious "you're safe in my hands, baby" attitude. Pete the airline captain is all authority and competence.

Around the world, tens of thousands of airline pilots make a similar transformation. That harried blonde in line at Starbucks, the guy pumping gas into his pickup truck—when these suburban parents and weekend anglers don their uniforms, airlines trust them with multimillion-dollar airplanes and the companies' reputations. This is why the making of airline pilots is serious business. Airlines screen them before they hire them, test them once they're in, and train them repeatedly throughout their careers. The airlines want seemingly contradictory qualities in their pilots: decisive but open-minded, vigilant but flexible, experienced but constantly learning, adherence to standard procedure but with an ability to improvise when required.

Creating a "pilot-type person" is so important that 95 percent of the cadets hired by Lufthansa have not flown at all when they arrive at the company-owned Airline Training Center in Arizona, fifty-seven hundred miles from company headquarters in Frankfurt, Germany. The thinking at the carrier is that if they can find people with the right personality, they can "grow their own pilots," as Matthias Kippenberg, president and CEO of the training center, explained. I saw this process up close in the fall of 2010, when I joined class number NFF380 for a week.

Lufthansa might as well be looking for astronauts; that's how hard it is to get selected. And with good reason: at the time, it cost thirty-five million dollars a year to run the school that turns cadets into airline pilots. "Students are selected by the airline, trained and sponsored by the airline," Kippenberg said. "They have a job guarantee," he said. "All they have to do now is learn to fly."

Kippenberg graduated from ATCA in 1977 and has been running the program since 2002. He may not have appreciated how hard his students work to get accepted until his own daughter Lisanne applied at the age of nineteen to take an Introduction to Aviation course offered by the Swiss government. She qualified to apply because her mother is Swiss, but, like everyone else, she would have to take a series of aptitude exams. In preparation, she and her dad went online for guidance—only to be confounded by the very first practice test. On the screen were six three-dimensional cubes. An X was placed on one of the six sides of each cube. Lisanne had to listen as a voice instructed her to imagine the cube rotating up or down, right or left, back or front. Lisanne's job was to keep track of where the X ended up.

"My dad and I were looking at it, and then we looked at each other; we didn't know what was going on," she told me. The "nine clocks test" was equally baffling. I'll spare you the details.

The students in my dorm at ATCA told me they'd performed similar tests, and others in which they had to listen to long series of numbers and repeat them back in reverse order. It sounded like torture to me, and I had the chance to experience it myself when CTC Wings of New Zealand put me through my very own pilot aptitude evaluation.

This differed from the kind of screening conducted by airlines because applicants at CTC are paying their own way, with no assurances they will find a piloting job. The aptitude test is necessarily less selective and less intense. Even so, having to keep the wings of my simulated aircraft straight and level while flying through a series of yellow rectangles appearing on my computer screen, I was tense and sweaty even before the mental processing exam began.

My overall score suggests that as a pilot, I make a great writer. I could process information okay and even acquire data under time pressure, judging by how quickly I was able to make sense of an image of shattered glass. (Don't ask.) When it came time to handling a joystick while following a flight path and working math questions involving counting backward, I was fried.

Viktor Oubaid, head of the German Aerospace Center in Hamburg, told me that all these challenges are designed to test working memory. One could practice and get better, as Lisanne and the Lufthansa students did, "but the maximum possible performance depends on your original abilities," Oubaid said. "In other words: many people can learn to fly, but only some are able to work as airline pilots."

After several months spent practicing for the tests, Lisanne went to Dübendorf, outside Zurich, for her entrance exam. The setting was quiet, and she had a good feeling. "I did a lot better than I thought I was going to do," she told me. Sure enough, Lisanne was accepted. After two weeks of flying lessons, she said she was even more excited about a career as a pilot and more appreciative of what the tests were trying to determine about her cognitive abilities.

Lisanne's dad knows very well this aspect of her experience. Call it multitasking or workflow management—pilot aptitude tests are designed to detect this and other things because so much more than epaulets is riding on the shoulders of the men and women on a flight deck. This was made most evident five years later, when one of the students at Kippenberg's school passed right through the airline's tightly woven web of diligence.

Andreas Lubitz arrived in Phoenix right after I'd left, when he was twenty-three years old. He had aced five days of testing and had interviewed with confidence back in 2008, when he was selected to begin ground school in Bremen, Germany. Yet Lufthansa's notably tough standards may have proven to be too much, because after two months, he took a leave of absence. From January to October of the following year, he underwent psychiatric treatment for reactive depression that a German medical examiner told the FAA had been triggered by excessive demands.

By 2010, Lubitz was considered fit to continue his training, and so he did: ground school in Bremen, and then flight school in Phoenix, followed by jet training back in Bremen and a stint as a flight attendant. In 2013 he became a first officer on Lufthansa's low-cost carrier GermanWings.

In the spring of 2015, Lubitz would commandeer his own flight from Barcelona to Dusseldorf and fly it into a mountain, killing himself and 149 others. The thirty-four-year-old captain, Patrick Sondenheimer, had left the cockpit to go to the bathroom after leveling the plane at thirty-eight thousand feet. With Sondenheimer gone and the cockpit door locked, Lubitz put the Airbus A320 on an autopilot descent to one hundred feet, a path that would take the plane directly into the high terrain in the French Alps.

Lubitz overrode the captain's attempts to return to the flight deck and did not reply to radio calls from controllers. For eleven minutes the plane descended, until finally it hit a mountain near Prads-Haute-Bléone. It would soon come out that Lubitz's depression had returned, and that in the weeks before the event, his physicians had advised him not to work, according to notes found in the trash in the young man's home.

Suicidal and/or homicidal airline pilots are a special kind of scary, even though this is an exceedingly rare occurrence. During its investigation into the crash, the French air accident bureau reported six similar events in which commercial pilots had deliberately crashed planes with passengers aboard. Japan Airlines in 1982, Royal Air Maroc in 1994, SilkAir in 1997, EgyptAir and Air Botswana both in 1999, and Mozambique Airlines Flight 470 in 2013. These crashes were all believed to have been purposely initiated by one of the pilots. With more than 717 million flights since 1980, you can see how infinitesimal this particular flight safety risk is.

It is an entirely different story when you get to unintentional acts. Then the mistakes pilots make that contribute to disasters are everywhere you look.

Sole Responsibility

None of the six people on the private jet from Samoa to Melbourne on November 18, 2009, will ever forget the night they ditched into the choppy waters of the Pacific off the coast of tiny Norfolk Island. Capt. Dominic James and First Officer Zoe Cupit were piloting the medical evacuation flight, with patient Bernie Currall, her husband, Gary, and a medical team riding in the back. As the Israel Aircraft Industries Westwind jet approached the island where the plane was to refuel for the last leg to Melbourne, the weather was so bad that neither James nor Cupit was able to see the runway. After four missed approaches, James decided to put the plane down in the ocean before it ran out of fuel.

The jet split in two on impact with the rough sea. Only three of the six people on board had life jackets, but all got out of the plane. For ninety minutes, James played shepherd, swimming around them in a circle and keeping them together. Finally, a search party on a charter fishing boat spotted the light from the small flashlight James was waving toward shore, and the group was rescued.

"It gives me goosebumps still thinking about it," said Glenn Robinson, an island resident and one of the crew members on the boat that rescued the survivors. "They're all alive. You know they've ditched that plane into a rolling ocean in the middle of the night, and here they are."

Brainy, articulate, tenacious, and the spitting image of the actor Tom Cruise, James was a hero and a celebrity when he arrived back in Australia. "Gold standard" is how he and Cupit were described by John Sharp, the chairman of Pel-Air, the company whose airplane the pilots were operating that night. By Christmas Eve, however, the two had fallen from grace. The Australian Civil Aviation Safety Authority (CASA) suspended their licenses, claiming the two had demonstrated bad airmanship. The accident was entirely the fault of the captain, said John McCormick, CASA's chief.

A man who puts a jet aircraft into a dark and stormy sea on a moonless night and then, without a life preserver, keeps his passengers together for an hour and a half like some kind of aquatic sheepdog is not the kind of man to allow himself to be made a scapegoat.

In planning for the flight from Apia, Samoa, James said he had been hampered at every turn. Unable to get WiFi on his phone or at the hotel, he did his flight plan on his cell phone in the hotel parking lot. He fueled up the jet with the assumption that it would fly in reduced vertical separation minima airspace (RVSM), a horizontal slice of the sky between twenty-eight thousand and forty thousand feet that requires planes with highly calibrated altimeters and special certification, both of which this Pel-Air plane did not have. Because flying in RVSM airspace can reduce fuel consumption, pilots routinely beg their way in by explaining that they are on a medical flight. James said it was a company practice he had complained about the year earlier, but nothing had changed.

When weather deteriorated on the way, James did not have enough fuel to make it to an alternate airport. He was not required to, under the work rules that applied to air ambulances. This was a loophole that national safety authorities had been trying to close for years, but CASA had failed to take action.

Years later, when James got his license back and started flying for other operators, he realized that many tools were available that could have changed the outcome that night. "I had access to flight planning software wherever I went. I had access to performance data, so I could look at a destination and calculate weights, speeds, and options," none of which he had access to while flying for Pel-Air.

These shortcomings and others were noted in a special audit of the airline that CASA conducted right after the ditching. Thirty-one safety deficiencies were found. CASA noted a conflict between "the commercial objectives of the company and safety outcomes." That's bureaucrat-speak for saying the operator was "more worried about profits than safety." Nevertheless, the Australian aviation authority's position was that the operation of the airline was not relevant. The accident, it stated, had been "caused by poor fuel planning, poor decision-making" by the captain.

Two years before the Pel-Air mess, George Snyder wrote in an article for the Flight Safety Foundation's *AeroSafety World* magazine, "The assignment of blame artificially and *prematurely* restricts the investigation process" and can even stop the investigation in its tracks. It was a prescient bit of writing, because that's exactly what happened in the Pel-Air case. Even knowing the airline had safety lapses, the head of CASA, John McCormick, did not share them with the accident investigators because the ditching, he said, "was entirely the fault of the captain."

Captain James admits he made mistakes, but adds that none of the support that would have helped him do his job was there. "I didn't operate in a vacuum. I operated as a pilot who belonged to a company that was overseen by a regulator," he said. "You can't isolate one from another and say that's a fair appreciation of what took place."

An Australian television documentary program called *Four Corners* presented the full story in 2012, just as the ATSB was issuing its probable cause report. The program prompted several parliamentary hearings into the way the aviation agencies were doing their jobs. This intrigued me, because issues of safety can be nuanced, and that isn't always appreciated in politics. Yet here was a case where the politicians sounded reasonable while the aviation professionals were looking no further than the pilot.

"It is surprising and dismaying both," said John Lauber, a research psychologist in the field of human factors—essentially, trying to understand why people do the things they do. He had been a member of the NTSB, and he'd spent most of his career trying to improve support systems for pilots. "All human performance takes place in a context, defined by the technology, procedures, and training."

On a bright morning in October 2014, my sister Lee and I were on vacation in Darwin, the capital of Australia's Northern Territory. We'd booked a seaplane flight into the aboriginal territory called Sweets Lagoon. Outside our hotel, the bus to the airport was waiting, and out of the driver's seat bounded a tanned and attractive man who looked vaguely familiar.

"Christine Negroni?" he asked with a smile, extending his hand. "Dom James." By then I had recognized him. My surprise was seeing him driving a tour bus. When he'd told me his story over lunch at a restaurant in Sydney half a year earlier, he was dashing off for a pickup job he had flying corporate jets. By then, CASA had restored his pilot's license. He was hoping for full-time work, but he had been vilified by the nation's top aviation official, and that had taken a toll. He would go for job interviews, but it was always the same story. "All these people would say, 'the crash, the crash, the crash.' They didn't know about the senate inquiry. All they know is I'm some guy who crashed a plane," he told me. "You can't make someone be educated on the nuances of the accident."

Pilots err for many reasons, Lauber told me. "To say a pilot made a bad decision is not a reflection on that pilot, but a reflection on the overall design of the system that he is tasked with operating."

The Prevention System

There's a story about an airline captain who, having landed at the airport and parked the jet at the gate, announced to the departing passengers, "Welcome to your destination, ladies and gentlemen. The safest part of your journey has come to an end."

Flying is so safe that we can appreciate the joke. From the very first plane crash in 1908, attention has been focused on finding out what went wrong and how to fix it. For decades this meant modifying the airplane or engines, and sometimes both, as was the case with the Comet; or coming to grips with new technology, as with the Dreamliner. Dana Schulze had spent more than a decade as an air safety investigator when she started the two-year project to understand what happened on the 787. She told me it was rare to have a case where her team was concentrating solely on the machine. By the time the probe was finished, though, human errors had been discovered in quality control during the manufacturing of the battery cells and in the assumptions made by engineers during the certification of the airplane.

Still, Schulze's expectation demonstrates the transition over time to more reliable airplanes and engines. They simply don't fail as they did in the early days of air travel. What has not changed is the fallible human, unpredictable at every level except for how reliably he or she will make mistakes. Something had to be done about making people perform better. Mechanical engineers and aerodynamicists had their place, but beginning in the 1970s, a new kind of specialist was digging into the "soft" sciences of psychology, ergonomics, communication, and design. These people were practicing in a relatively new field called human factors. They worked on ways to help airlines select and train pilots. Programs widely used in the military were adapted for civilian airlines. They researched how to enhance the flight deck and improve communications—what, in the jargon of aviation, is called information transfer—so that misunderstandings could be averted and errors prevented before they ended in tears. When a disaster happened, it could be turned into a valuable learning opportunity. The three biggest developments in human factors were triggered by pilot error, including a development as simple as it is profound: the checklist.

In 1935 three airplane manufacturers were competing to provide the U.S. Army Air Corps with a bomber capable of carrying a ton of ordnance a distance of two thousand miles. Boeing thought it had just the thing: an all-metal, four-engine model it called the Boeing 299. A *Seattle Times* reporter named Richard Williams, upon seeing the enormous new airplane for the first time, reportedly exclaimed, "Why, it's a flying fortress!" and the nickname stuck.

Douglas and Martin were also in the running with twin-engine models. The Martin 146 and the Douglas 1B could carry the load, but neither could go the range. Even before the flying demonstration began, the army procurement officers were pushing to buy sixty-five Flying Fortresses. The competition was Boeing's to lose.

On October 30, the day of the fly-off at Wright Field in Dayton, Ohio, two army pilots, two men from Boeing, and a representative of the engine manufacturer Pratt & Whitney climbed in and took off. There were three pilots aboard, Ployer P. Hill and Donald Putt from the army, and Boeing's chief test pilot, Leslie Tower. Yet as they taxied across the airfield, none of them noticed that the elevators and rudder, movable panels that control the plane's up-and-down and side-to-side motion, were still secured with a gust lock that kept them in place so they did not swing in the wind and get damaged when the plane was on the ground. But when the plane took off it was locked into a configuration for a steep climb that the pilots could not correct in the air. The giant aircraft stalled and fell to the ground, erupting in flames. Two of the men died from their injuries.

Before that flight, the only negative about the Boeing 299 was that it might be too complex, but it was a simple oversight that brought it down, literally and figuratively. The plane was disqualified from consideration by the Army Air Corps, and the big bomber order went to Douglas.

What could be done to protect against forgetfulness? The army pilots got together and came up with a checklist. In twenty-four steps, from "before taxi" to "after takeoff," the pilot would be reminded of each critical task. Boeing went on to build the B-299, and the Army Air Corps bought it and flew it for decades. The Flying Fortress entered the history books. So, too, did the pilot checklist.

Checklists aren't a perfect solution. Every fix has unintended consequences. The same checklist read eight times a day might not be met with the same level of concentration, a phenomenon that former NASA scientist and human factors expert Dr. Key Dismukes calls "seeing but not seeing."

The number of checklists on an uneventful journey is about a dozen. On an eventful one, such as Qantas Flight 32, an Airbus A380 jumbo jet that experienced an uncontained engine failure shortly after takeoff in 2010, the pilots went through about one hundred twenty checks. Due to the way the Rolls-Royce Trent 900 engine exploded, Capt. Richard de Crespigny and his first officer, Matt Hicks, had one engine out; the three others not working properly; an inability to transfer fuel; problems with electrics, communication, flight controls, hydraulics, and pneumatics; and "a whole bunch of other stuff," as de Crespigny described it.

They were busy assessing the condition of the plane and planning their next steps while emergency checklists kept appearing on the flight display "like dinner plates at an all-you-can-eat buffet," according to de Crespigny. "I think I invented the term checklist fatigue," he said—but he hadn't. De Crespigny just had the most high-profile experience with the phenomenon of checklists overwhelming pilots, characterized as "stop interrupting me when I'm busy" and identified by Dismukes back in 1993.

Dismukes wasn't calling for an end to checklists, which, if used properly, might have prevented disasters such as Helios Flight 522, where pilots seem to have neglected to turn on the cabin pressurization switch; or the Spanair Flight 5022 crash that followed an attempted takeoff without flaps. He was recognizing that as machines grow in complexity, every part of how humans interact with them must evolve, too. "How complex is too complex?" is a question that would come up again and again.

In the summer of 2009, the sixty-year-old captain of a Continental Boeing 777 keeled over at the controls of a flight carrying 247 people from Brussels to Newark. Passengers heard flight attendants ask if there was a doctor on board. There was, but it was too late for Craig Lenell. All airlines are required to have two pilots, and on flights over a certain number of hours, there can be three or more: the crew flying and a pilot or crew in reserve.

"If something happens physiologically to one of the pilots, the other one is seamlessly able to carry on," ABC News aviation consultant and retired airline pilot John Nance said.

A pilot dying at the controls is pretty rare. The benefit of having two pilots is realized far more often in less dramatic circumstances. As Nance explained, two pilots provide "two brains and two sets of eyes" for the flight. At its best, a two-pilot crew operates as a precision team sharing a common view of the task before them, and separate views of each other. The terms *challenge and response* and *monitoring and cross-check* and the information transfer I mentioned earlier describe this relationship, which might also be called communication.

Entire books have been written on how to improve the way pilots communicate, because without special training, pilots can misunderstand each other as easily as any other two people who meet briefly and maybe for the first time and set out to accomplish a task together. Yet in the cockpit, the stakes are higher.

The deadliest aviation accident ever, the collision of a KLM Royal Dutch Boeing 747 with another 747 being operated by Pan American World Airways, made it very clear that more attention needed to be paid to the seemingly simple task of talking and the complex phenomenon of hierarchy on the flight deck. The crash happened on March 27, 1977, on a runway in Tenerife, in the Canary Islands.

Both planes and several others had been diverted from Las Palmas in Gran Canaria to the Los Rodeos Airport in Tenerife. A bomb at the Las Palmas airport terminal had shut it down, and three hours passed before it was reopened. When it was, airline crews started preparing to leave Tenerife for the short flight to their original destination, Las Palmas.

On the KLM plane the pilots in the cockpit were under the command of fifty-year-old Jacob van Zanten, head of flight training and, as the face on the airline magazine ads, something of a superstar in the Netherlands. The delay must have weighed on van Zanten, who was operating along with the other pilots under new flight time restrictions. If they were held up much longer, they would not be allowed to make the return flight from Las Palmas to Amsterdam.

If that happened, hundreds of travelers would have to be accommodated in hotels, and the jumbo jet would sit at the airport overnight instead of providing revenue to the carrier. And there was another factor to consider: the commander's ego. As Nance explained, there is "the embarrassment of a senior leader in being unable to make happen what he wanted to happen." In the years to come, this and other similarly subtle pressures would be explored more thoroughly for their impact on the decisions made that fateful afternoon.

There were far more planes at Los Rodeos than gates at which to park them, so planes parked on a few of the taxiways. But this created a new problem: they were blocking the way to the runway for departing flights.

Controllers told the departing crews to follow an unusual procedure known as backtaxiing. One plane was to taxi down the runway followed by the second. When the first plane arrived at the departure threshold, it would make a one-hundred-eighty-degree turn into takeoff position.

The following plane had to pull off onto a taxiway to get out of the way of the plane positioned for takeoff.

When the first plane departed the second crew would taxi their plane into position and go.

Two planes had already taken off, and Pan Am and the KLM wide-bodies were next. KLM led the way, followed by *Clipper Victor*, under the command of the coincidentally named Victor Grubbs, fifty-six. Robert Bragg was the first officer and George Warns was the flight engineer.

Rumbling down the runway, Bragg, 39 at the time, recalled that the skies were clear, but before the Pan Am plane could find the taxiway where it was to pull off to get out of the KLM's takeoff path a dense fog rolled in. "Our visibility went from unlimited to 500 meters in under one minute. The tower even made a call stating, 'Gentlemen, be advised that runway visibility is 500 to 700 metres,'" Bragg, thirty-nine, wrote in an article for *Flight Safety Australia*. The fog was so thick that the Pan Am crew determined it was below the takeoff minimum and assumed that the runway was now closed. Grubbs, a pilot with twenty-one thousand flight hours, continued to steer the plane through the fog, but slowly. All three men strained to find the turnoff.

On the takeoff end of the runway, however, the fog had passed. Van Zanten pivoted the KLM airplane around to point in the direction of takeoff. His plane was now head to head with the slow-moving Pan Am jumbo, unseen in the cloud a half mile ahead. He pushed forward on the throttles—which seemed to startle his first officer, thirty-two-year-old Klaas Meurs. "Wait a minute, we don't have ATC clearance," Meurs said. The first officer had experience with this captain. It was van Zanten who had qualified him on the 747 just two months earlier.

Van Zanten pulled back the power and instructed Meurs to call ATC. The first officer radioed that he was ready for takeoff and waiting for clearance. The tower controller replied with departure and navigational information, but didn't issue the clearance to take off.

Just as Meurs was confirming the instructions, van Zanten said, "We go, check thrust." Once again, the captain fed fuel to the airplane's engines.

Meurs cued the mic and read back the controller's words, adding something that sounded like "We are now at takeoff."

In an analysis of the accident by the ALPA, this ambiguous phrase would receive a lot of scrutiny. The pilots concluded that the KLM first officer thought something was wrong with Captain van Zanten's decision and was "trying to alert everyone on frequency that they were commencing takeoff."

Captain Grubbs, on the Pan Am Clipper, heard the transmission and was surprised. "No," he said, followed a second later by the controller telling KLM, "Stand by for takeoff. I will call you."

"And we are still taxiing down the runway," Pan Am's First Officer Bragg added.

The Pan Am pilots must have thought they were making it

clear to KLM that the Pan Am 747 was still in the way, but these messages were drowned out by the dueling transmissions, which created a shrill noise in the KLM cockpit.

"Report when runway clear," the controller told the Pan Am crew, and Pan Am responded, "Okay, we will report when clear."

The KLM 747 was accelerating down the runway; its first officer, Meurs, and flight engineer, Willem Schreuder, heard this. It prompted Schreuder to ask, "Is he not clear, then?"

"What do you say?" Captain van Zanten asked.

"Is he not clear, then, that Pan American?" Schreuder repeated.

"Oh yes," van Zanten and Meurs answered at the same time.

Captain van Zanten had made his decision. "To reassess that decision at such a critical point in the takeoff may have seemed an intolerable idea," the human factors specialists at ALPA concluded in their report, citing the other factors that must have been on the captain's mind, including the heavy airplane, wet runway, and poor visibility.

The KLM Boeing 747 was closing in on the other jumbo jet still obscured in the soup.

On *Clipper Victor*, Captain Grubbs was startled to see the lights of the KLM airliner rapidly approaching through the mist.

"Goddamn, that son of a bitch is coming straight at us," he said. He applied power to the throttles and turned the nose wheel to the left in a desperate attempt to get out of the way.

At this point, KLM's van Zanten also saw the impending collision. With his aircraft moving too fast to stop, he pulled back on the yoke in a frantic effort to take off over the Pan Am Clipper. The tail of the KLM plane scraped along the runway as the front end lifted enough for the nose gear to clear the dome of the Pan Am 747.

Having been seated on the right side of the Pan Am flight deck facing directly toward the KLM jet, First Officer Bragg remembers his horror. "Get off, get off, get off," he screamed at Grubbs as the underside of the KLM plane rose above him.

"I ducked, closed my eyes, and prayed, 'God, let him miss us.' When it did hit our plane, it was only a very short, quiet shudder. I actually thought that he had, in fact, missed us until I opened my eyes."

As the KLM 747 dragged its undercarriage across the upper deck of the *Clipper Victor*, it tore off a huge section and then slammed back down onto the runway, skidding another fifteen hundred feet in a burst of sparks and explosions as the fuel spraying from ruptured tanks ignited.

All 249 people aboard the KLM flight were killed. Of the 396 people on the Pan Am jet, 70 survived the crash (though nine died later), a fact Bragg credits to Grubb's quick work getting at least the front end of the plane out of the way.

Among the survivors, some said the impact was like a bomb going off. Its effect on the airline industry was equally explosive. It was not that two airliners had collided; there had been others. But the number of casualties and the cascading series of communication failures was a loud wakeup call to the industry.

Everywhere one looked, errors had been made, most obviously by the three men on the KLM flight deck, who failed to communicate their concerns clearly. The collision also revealed the danger of aviation's long-established "right stuff" religion. The dogma consists of the belief that the captain is always right and that good pilots never make mistakes. In a 1990 article for the Flight Safety Foundation, Robert Besco, a retired airline captain and a consultant in human performance, wrote, "Pilots have adopted an attitude of risk denial." If the captain was considered God and everyone else a congregant, you can see how pilots would not/could not speak up even when they saw that something was wrong.

John Lauber, working at NASA's Ames Research Center in California at the time, had already spent several years noting the growing disparity between the reliability of the machine and that of the human flying it. He had visited airlines and spoken about a new concept he called cockpit resource management, or CRM. One of the airlines he visited was KLM Royal Dutch. One of the pilots he met was Captain van Zanten.

Lauber remembered, "He was a very impressive guy, a blond, steely-eyed airline pilot. He was a strong-minded personality." Lauber was pitching CRM as something airlines could use to train pilots to better manage their workplace. Yet they "had not done anything in terms of developing programs that address these issues," Lauber recalled.

CRM was about more than teaching communication and moderating hierarchy. It was intended to help pilots manage a wide array of pilot errors that Lauber had seen in a review of eighty accidents in the 1970s and '80s. The one that stood out was the flight into terrain of a brand-new Lockheed L-1011 in Miami in 1972.

Capt. Bob Loft was one of Eastern Airlines' most senior captains, with thirty thousand hours of experience. His first officer, former air force pilot Bert Stockstill, had six thousand hours and even more time in the L-1011 than Loft had. The flight engineer, Don Repo, was a twenty-five-year employee with Eastern with nearly sixteen thousand flight hours.

As Eastern Airlines Flight 401 approached Miami International Airport in late December 1972, the nose wheel landing gear indicator light did not illuminate. Without knowing whether the failure was of the light or of the gear, the captain did a go-around. Cleared to fly at two thousand feet, the crew began diagnosing the situation, and during the process of removing the light fixture and trying to reinsert it, someone inadvertently turned off the altitude hold. This went unobserved by the three pilots and the mechanic occupying the jump seat because all four men were trying to figure out whether the problem was with the gear or just the warning light.

As the plane dropped from its assigned altitude, an alert began to sound, but no one seemed to hear it. They were close to concluding that it was just a bad bulb when Stockstill noticed the plane's descent.

"We did something to the altitude," the first officer said.

"What?" said Captain Loft.

"We're still at two thousand, right?"

Loft's final words showed his confusion: "Hey, what's happening here?"

Flight 401 smashed into the Everglades, killing 111 of the 177 on board.

When John Lauber came across the detailed report he called it a "prototype" of an accident in which the crew does not manage the resources available. His cockpit resource management would teach pilots how to do this, in the same way businesses train their managers. "Pilots generally were well trained on aircraft systems and basic flying skills," he said. But nothing was done to teach them what they needed to know for decision making, communication, and leadership.

The Tenerife accident gave Lauber's work new energy, and in the years to come, cockpit resource management would be changed to "crew resource management," in recognition that other flight personnel such as mechanics, flight attendants, dispatchers, and air traffic controllers had a role to play in safe flights. As a bonus, the acronym, CRM, remained the same.

"So many incidents in life as well as [in] other industries have broken down because of the ambiguity in communications," said Christopher D. Wickens, a professor of psychology specializing in aviation human factors. "CRM is clearly one of the most important things that has developed in aviation over the past forty years."

CRM was sometimes dismissed by those who said it's a sissy notion, an "I'm okay, you're okay," touchy-feely exercise. Overall, resistance has subsided, though there remains a challenge in eliminating what Wickens calls the negative authority gradient, when differences in rank and experience in the cockpit create communication difficulties.

The power divide still restrains lower-ranked or less-experienced pilots from calling errors to the attention of a senior pilot. In a report for NASA, Dismukes and fellow human factors scientist Ben Berman discovered that captains would correct copilots when they made mistakes twice as often as first officers would when they saw the captain err.

"The cockpit traditionally was a strict hierarchy; the junior pilot never asked questions. Part of CRM training is to create an environment that, when [the junior pilot] has information that's critical to the flight, the captain will listen," Wickens said. Drawing a parallel to how rank is disregarded in safety-critical situations in the military, Wickens explained, "Landing on aircraft carriers, a low-ranking person can be in charge of things because they have the information that everyone needs. Authority shifts dynamically."

Pointing out errors can make for difficult conversations, and many inhibiting factors were in play on the night in 2009 when a Colgan Air turboprop crashed on approach to the airport in Buffalo, New York. Rebecca Shaw, the twenty-four-year-old first officer, had been flying for Colgan for one year. Marvin Renslow, the forty-seven-year-old captain, had four years with the company but only a hundred hours flying as a captain on the Bombardier Q400. While Shaw sniffled with a head cold and responded with a lot of "uh-huhs," the captain kept up a nearly one-way dialogue, even on approach to the Buffalo airport. Whether the first officer considered the banter a distraction isn't clear. She did seem worried about the difficult conditions in which they were flying: at night, in ice. A reading of the CVR suggests she was not inclined to assert herself. Even her apprehension about the ice was less than direct: "I've never seen icing conditions. I've never deiced. I've never seen any . . . I've never experienced any of that." She continued: "I'd have, like, seen this much ice and thought, 'Oh my gosh we were going to crash.'"

As the plane neared the airport, Renslow mishandled a stick shaker alert that the plane was flying too slow, presumably because it had accumulated ice, though in actuality it had not. Stalling protections on the plane caused the nose to go down to gain airspeed but the pilot pulled it back up, exacerbating the problem. The plane crashed into a house near the airport, killing everyone on board and one person on the ground.

It was an entirely different accident in terms of specifics, but a case of the same reticence to speak up, when Asiana Airlines Flight 214 landed short of the runway at San Francisco International Airport on a clear summer day in 2013. A series of misunderstandings about the way the automation worked meant that the flight was coming in too low and too slow, and the decision to go around and try the landing again came too late.

The plane hit a seawall at the edge of the runway bordering San Francisco Bay, slammed onto the ground, and pivoted up before hitting the runway a second time. Lee Kang-guk, the captain, in the left seat, had ten thousand total flight hours on other jets but just thirty-three on the Boeing 777. He was transitioning from the Airbus A320 narrow-body under the supervision of Capt. Lee Jung-min, who was in the right seat.

After the accident, Lee Kang-guk told investigators that he delayed initiating a go-around because he thought "only the instructor captain had the authority."

How open pilots are to asserting themselves, pointing out the errors of superiors, or acknowledging their own fallibility is highly influenced by culture. In an analysis in the *Journal of Air Transportation* in 2000, Michael Engle wrote that "there were extreme cultural differences" about whether "junior crewmembers should question the actions of captains" depending on where in the world they were from.

Forty years after the push to improve cockpit interaction and imbue the entire flight crew with a sense of shared responsibility, it seems the techniques work better in societies where individuality is valued more than rank. CRM may need to evolve to take into consideration the vastly different standards people have about interpersonal communication in parts of the world where aviation is experiencing the strongest growth, as in Asia, the Middle East, and South America.

The crashes of Colgan 3407 and Asiana 214 also shine a light on a hydra of issues that arrived like stowaways on the digital airplane: automation, complexity, and complacency.

Evolution

In the early days of flight, the cockpit was a busy and crowded place. The crew complement on the *Hawaii Clipper* in 1938 consisted of captain, first officer, second officer, third officer, fourth officer, engineer, assistant engineer, and radio operator—eight people required to fly six passengers. Each new generation of airplane incorporated advances that did better and faster a task formerly accomplished by the pilot. To fly as Wright had done meant to operate the machine with one's body and engage with one's senses. Each new advance made piloting less physical and more cerebral.

A normal crew consisted of three when Robert Pearson got his first job as an airline pilot in 1957. He was a first officer on the DC-3 for Trans-Canada Air Lines, which would become Air Canada in 1965. After flying the four-engine British Vickers Viscount, the DC-9, and the Boeing 727, Pearson was a forty-seven-year-old captain in 1983 when Air Canada went out and bought the world's most modern jetliner, the two-engine wide-body Boeing 767. This airplane was radically different because of the incorporation of technology that eliminated the need for a third pilot. The flight engineer (sometimes called the second officer) had been responsible for supervising the airplane's fuel, hydraulics, pneumatics, and electrical systems. But with the computers on the 767, the plane could monitor itself and present all that information to the pilots in bright, graphic, easy-to-read flight management system monitors.

The Boeing 767 was one step ahead of Airbus, which was producing an even more radical airplane, the first-generation fly-by-wire airliner that would put a computer between the flight controls and the control surfaces and create a protective flight envelope outside of which the pilot could not fly.

The digitization of flight started a new era, but were the airlines ready?

In February 1983, Pearson began a four-week course to qualify on the 767: two weeks of ground school and two weeks of flying the simulator. By April he was a captain. Sitting in the left seat, gazing at the array of gadgetry, he noted how many manual functions were now handled by the computer. "What did I know about computers? My experience with computers was using a Royal Bank of Canada ATM," he said. He was about to have a near-catastrophic experience on the Boeing 767, the origin of which was in not understanding the basics of the new airplane's technology.

On July 23, 1983, Pearson and First Officer Maurice Quintal were assigned to fly one of Air Canada's new 767s from Montreal to Edmonton. Due to a series of misunderstandings, the ground crew calculated the amount of fuel to load on the airplane by converting fuel volume to pounds, which is how they filled the other airplanes in the Air Canada fleet. But the 767's fuel system used kilograms. Since a pound is less than half a kilo, the error meant that only half the required fuel was pumped into the tanks for the four-and-a-half-hour transcontinental flight. The fuel quantity display was not working, so the crew manually entered the number 22,300 into the flight computer—without realizing that the plane's computer would consider it 22,300 kilos, or twice as much fuel as it actually contained. With the crew thinking the aircraft had enough fuel for the journey and then some, the plane departed.

Neither the pilots nor the fuelers realized their error, and the 767 no longer had a flight engineer managing the system whose job it would have been to ensure that the plane had the correct amount of fuel for the journey. "If everyone is trained and the lines are drawn as to who is responsible for what, there's no ambiguity," said Rick Dion, an executive with Air Canada maintenance who was a passenger on the flight. "In this case it was sort of open-ended. We weren't aware who was responsible for the final say on this fuel stuff."

Flight 143 was flying at forty-one thousand feet, about one hundred miles short of Winnipeg, when the first engine ran out of fuel, followed closely by the second. Without the engines to generate power, the pilots lost their flight deck instruments. They were seventy-five miles from the nearest airport. The riveting story of how experience and teamwork saved the day follows in part 5 of this book. The lesson here comes from Pearson, who said he and others learned that day that they were unprepared for the monumental leap in technology—and this from a man who had literally flown into the jet age.

"Transitioning from the noncomputer age to the computer age was more difficult than transitioning from propeller planes to jets, and it wasn't because they flew twice as high and twice as fast. It was all the big unknowns," he said.

After years of accidents attributable to pilot error, automating some functions was intended to make flying more precise, more efficient, and of course safer. A look at the decline in the rate of air accidents since the arrival of the digital airplane shows the benefits. The number of crashes resulting in the loss of the airplane, known as a "hull loss," has remained stable over the years, while the number of flights increased from half a million a year in 1960 to nearly thirty million in 2013. The third and fourth generation of automated airplanes, those with digital displays and computers that protect the airplane from maneuvers outside a predetermined range of safe flight parameters, are even more effective.

Automation's downside is that it creates both complexity and complacency. The complexity can cause pilots to misunderstand what the airplane is doing or how it works. It was complexity that caused half a dozen Air Canada employees to be unable to calculate how much fuel to pump into Flight 143. It was the opacity of the system that led the pilots to think that by entering the amount of fuel they thought had been loaded into the tanks, they would get an accurate reading of the fuel available for their flight. Recognizing the mistake afterward, Pearson said he understood for the first time the expression "Garbage in, garbage out."

Considering how automation can lead to confusion, it is a paradox that it can also contribute to crew obliviousness. With the L-1011 on autopilot, all three men on Eastern Flight 401 turned their attention away from the controls to work on changing a lightbulb. More recently, a Northwest Airlines flight from San Diego to Minneapolis made headlines around the world when the pilots got so wrapped up working on their laptops that they flew past their destination.

Flight 188 was one hundred fifty miles beyond Minneapolis International Airport with 144 passengers in October 2009 when a flight attendant called the pilots, curious to know why the plane had not begun its descent. For fifty-five minutes, the pilots had failed to acknowledge radio calls from air traffic control in Denver and Minneapolis or calls from the flight crew of another Northwest plane. To this day, people suggest that the pilots must have fallen asleep, because how else could they have missed hearing all the people calling them on the radio?

Robert Sumwalt was a member of the NTSB at the time. A former airline pilot, he was familiar with the troublesome issue of complacency. In 1997, Sumwalt and two others went through anonymous pilot reports and found that failure to adequately monitor what the airplane was doing was a factor in one-half to three-quarters of air safety events. Between 2005 and 2008 an airline industry group found sixteen cases similar to Northwest Flight 188, including one in which a captain returned from the bathroom and found the first officer engaged in a conversation with the flight attendant. The copilot's back was to the instruments, so he did not notice that the autopilot had disconnected and the plane was in danger of stalling. After losing four thousand feet of altitude, the captain was able to recover control of the airplane.

When Flight 188 made headlines, then-FAA administrator Randy Babbitt got on the evening news and castigated the flight crew. He pointed out that the Northwest pilots were on their laptops doing work unrelated to the flight, a prohibited activity. "It doesn't have anything to do with automation. Any opportunity for distraction doesn't have any business in the cockpit. Your focus should be on flying the airplane."

Tough talk sounds good, especially when stories such as

that of the Northwest Flight 188 get blasted all over the news, making air travelers nervous. Still, telling pilots to pay closer attention is too simple. It may not even be possible to give unrelenting focus to routine tasks, according to Missy Cummings, a systems engineer and director of Duke University's Humans and Autonomy Lab. "The human mind craves stimulation," she said. Failing to find it, the mind will wander.

Cummings, a former navy F-18 pilot, is a proponent of automation, and envisions a future with more of it, not less, if the problems identified by one of her former students at Massachusetts Institute of Technology can be resolved.

While working on her masters at MIT, Christin Hart Mastracchio conducted a study that showed that when automation reduces a workload too much, vigilance suffers. "Boredom produces negative effects on morale, performance, and quality of work," she found. Now an air force captain at Minot Air Force Base in North Dakota, Mastracchio is a pilot on the sixty-year-old, eight-engine B-52 Stratofortress, and automation is not her problem.

"The B-52 is on the opposite end of automation. It takes five people just to fly it," she told me. "It takes all of us working together to control the monstrosity. You need to find a center point where you have the right amount of automation."

On the day that Asiana 214's Lee Kang-guk was making his first approach to the San Francisco airport while training to be a captain on the Boeing 777, an electronic navigational aid that would normally have been used was down for maintenance. Since July 6, 2013, was a clear, sunny day, this might not have been a problem for many pilots, but it's the practice of some airlines, including Asiana, to use automation all the time.

After the plane's crash landing, Capt. Lee Kang-guk told investigators he'd found making a visual approach "very stressful." It was "difficult to perform," he said, in the absence of the electronic system that tracks a plane's glide path.

Capt. Lee Kang-guk was an experienced captain on the Airbus, though it is important to remember that he had just thirty-three hours on the Boeing 777. On approach, he made a series of errors while trying to get the airplane on the glide path to the airport. Neither he nor the pilot supervising him, Capt. Lee Jung-min, even discussed doing a go-around despite the fact that company policy required it when a plane was not at the appropriate height or speed approaching five hundred feet. In one respect, Capt. Lee Jung-min, with twelve thousand hours, was like Lee Kang-guk: it was his first flight as a training captain.

All these factors and others played a part in the accident. In its report, the NTSB concluded, "More opportunity to manually fly the 777 during training" would help pilots perform better.

In the process of writing this book, I had the chance to listen to a familiar story told from a new perspective. The tale begins the week prior to the historic flight of Orville and Wilbur Wright. Samuel Langley, the head of the Smithsonian Institution, had been given a government grant of $50,000 (equivalent to $1.2 million in 2016) to develop a powered airplane. Throughout the summer of 1903 he had been tinkering with this one-man contraption called the Great Aerodrome. It was to be catapulted from a track mounted on a houseboat in the Potomac River, but prototype flights had not gone well.

On December 8, 1903, with Langley's assistant, test pilot Charles Manly, on board, the Great Aerodrome was pushed from the top of the boat, but it never became airborne. It collapsed and fell into the icy waters of the river. Discouraged, Langley gave up. Sixty-nine at the time, he may never have expected to live long enough to see man fly. Yet nine days later, the Wright Brothers made history with a twelve-second controlled flight at Kitty Hawk in North Carolina. That's the story I knew.

The goose bump–inspiring and brilliant postscript was presented to me by John Flach, a professor and chair of the Department of Psychology at Wright State University in Dayton, Ohio: "Christine, the Wright Brothers learned that for a plane to work, it had to put control in the hands of humans. That's a metaphor."

It is a metaphor appropriate for aviation's first century. But what about the second? Flying has gone from the days of the Wright Brothers controlling the plane by shifting their weight to pilots who sit at keyboards typing instructions that command a complex system of computers. For a while the debate has been over who or what does the job better, the human or the machine. What is emerging is that each does the job differently

"The computer is a rule-based system," Flach told me. "What it means to be reasonable and human is to break the rules. A computer will continue to do its computing while the building burns around it. A human will adapt to the situation."

The aviation industry has spent decades creating support for the stresses that pilots encounter, from choosing the right candidates to teaching them how to manage resources. And on almost every flight, new technology and age-old human qualities mesh in just the right way so that flying is safer than the sum of these parts.

PART FIVE

Resiliency

Every day, somewhere in the air, a cockpit crew averts disaster by routinely dealing with equipment malfunctions, weather uncertainties, or unscripted situations.

—DR. KEY DISMUKES,
NASA HUMAN FACTORS SCIENTIST

The Control Metaphor

The scenario of MH-370's disappearance that I describe in this book is a tragic illustration of Flach's metaphorical look at the Wright Brothers flight: control had to be in the hands of the pilot. On MH-370, a rapid decompression triggered a chain of events that required human control, but hypoxia may have stripped the pilots of their mental ability.

Fallibility sometimes leads to disaster. Far more often, however, resiliency saves the day. Pilots interrupt errors and correct oversights. They find workarounds and reshuffle priorities. They avert problems they don't even know they have—and not just once in a while, but on nearly every flight, without the passengers even knowing something is amiss.

On occasion, however, the malfunction becomes obvious.

Alarms started blaring on the flight deck of Malaysia Flight 124 eighteen minutes after takeoff from Perth Airport on August 1, 2005. The plane was climbing through thirty-eight thousand feet when Capt. Norhisham Kassim and First Officer Caleb Foong were startled by two very conflicting warning horns. The first alerted them that the Boeing 777 was flying too slowly and was in danger of stalling; the next indicated that the plane was flying too fast. Before either man had time to react, they were thrust back in their seats as the front end of the airliner abruptly turned up.

One experienced 777 captain told me that it had to have been a "WTF moment." Yet with great understatement, Norhisham recalled that he was "startled." The nose was high, about seventeen degrees, and rocketing farther skyward at a speed of ten thousand feet per minute. That's an ascent of more than one hundred miles per hour.

They had been flying on autopilot when the trouble started so Norhisham shut off the autopilot and pushed the yoke to get the nose down. That caused the autothrottle to rev, pumping more fuel into the engines and giving the descent an unexpected kick so that the plane started a four-thousand-foot dive. Moving the throttles to idle only made the plane bolt upward again, and the aircraft climbed two thousand feet before the confounded pilots could get it to stop.

The plane was shuddering violently, according to Kim Holst, a passenger from Australia. "A flight attendant dropped an entire tray of drinks and began crawling on his hands and knees back to his seat, and the other flight attendant began praying," Holst remembered.

For the pilots, "It was somewhat like riding a bucking horse," Norhisham said, adding, "As a passenger at the back-most seats, you definitely will feel worse than that."

Norhisham and Foong were facing a situation similar to Robert Pearson's experience on the Air Canada 767: a "garbage in, garbage out" encounter with the airplane's computerized brain.

All the fancy electronic gizmos feeding information to the pilots and enabling automated flight rely on sensors in the plane's air data inertial reference unit, called the ADIRU (pronounced ADD-uh-roo). The ADIRU consists of two sets of three accelerometers. Two accelerometers calculate the side-to-side/up-and-down movement of the wings called roll, two calculate the up-and-down motion of the nose called pitch, and two gauge the side-to-side/front-to-back motion on a horizontal plane called yaw. There are two sets of sensors for each; one is primary, and the second set is a backup. That redundancy provided such a feeling of security to operators of the Boeing 777 that no checklist was developed for what the pilots should do if there was bad information from the sensors. Yet this was what was causing MH-124's wild ride over Australia.

The crew knew that the plane was performing erratically, but not why. So as they prepared for a return to Perth, Norhisham hesitated to turn off the autothrottle. He had a handful of airplane, and he was hoping some part of the automation would help reduce the workload.

Like a bad driver pumping the accelerator, the lever controlling fuel to the engines kept "hunting up and down" as the two pilots struggled to get control of the flight. Norhisham and Foong were engaged in a kind of triage, dealing with the most critical problems as they nursed the 777 back to Perth. Faced with a confounding situation, Captain Norhisham had to give up diagnosing and concentrate on learning how to fly this plane with all its sudden idiosyncrasies.

Certainly they were relieved to see the approach to the airport, but that feeling was short-lived. They were about to get a last-minute surprise. As the plane descended through three thousand feet, another instrument in the orchestra of alarms began to sound. *Wind shear, wind shear*, the computerized voice called out, accompanying the shrill series of beeps.

Pilots are appropriately cautious about wind shear, which is a sudden change in the direction or speed of the wind. It is concerning close to the ground because it can cause a plane to lose lift, with little time for the pilot to recover. On the crippled Malaysia jet, the pilots were loath to get into a situation where they would have to rely on the plane performing as expected, considering how unpredictable it had been so far. Going around and trying the landing again, with all the uncertainties about the condition of the plane, also seemed risky.

Norhisham was worried. Was the alert real or another unreliable result from unreliable inputs to the ADIRU? The captain had to make a decision. The day was clear, visibility was good, and the wind was manageable, but the sun was going down. "We continued [toward the landing] with full caution," he said, keeping an eye out for any indication of wind shear.

"Thank God," Norhisham said when Flight 124 landed safely with no injuries, though everyone on board the airplane was shaken. Only then did Norhisham stop and think about the "very thin margin of survival." He had joined a fraternity of pilots who had knowingly broken the last link in the chain to calamity.

Three and a half years later, Chesley Sullenberger and Jeff Skiles ditched an Airbus A320 in New York's Hudson River after geese flew into the engines following takeoff from LaGuardia Airport.

In September 2010, Andrei Lamanov and Yevgeny Novoselov landed on an abandoned runway in northwestern Russia that was half as long as their aircraft required. A total power failure on a the Tupelov TU-154 caused all the fuel pumps to fail, starving the engines and leading to the loss of all navigation and radio equipment on what should have been a five-hour flight to Moscow.

Praising the actions of pilots like these, a writer for *New York* magazine called them "a dying breed." I don't think that's true. It is not that few can do what Norhisham and Foong, Sully and Skiles, Lamanov and Novoselov, did. It's that aviation's safety net is so expansive and robust that it is a blessedly rare occurrence when the pilots' full complement of talent, skills, and training is not enough to keep them from falling through it.

When pilots fail, it is headline news. When they succeed in addressing minor issues before they become major, however, it is for the most part invisible, making human resiliency the most mysterious of the many contributors to the industry's stellar safety record.

It is easy to see when things go wrong, writes James Reason in his book *The Human Contribution*. A professor of psychology and a pioneer in the study of human factors, Reason spent most of his career writing about why people screw up. At a conference in 2009, however, he presented what he considers the much more interesting flip side. Speaking to the annual Risky Business Conference in London, Reason called the subject of his life's work "human as hazard," a tedious subject. "It's banal; it's so everyday," he said. The octogenarian's attention had been turned to "the stuff that legends are made of," the qualities that allow humans to be heroes.

So much time has been spent learning from failure; what can be gleaned from studying the right stuff? This is not just Reason's question. From the icy day Sullenberger and Skiles ditched their plane in the Hudson, the public has clamored to hear their story. The same is true of the heroes who preceded them. The controlled crash landing of United Flight 232 occurred in 1989 and has been the subject of books and movies. The latest, Laurence Gonzales's *Flight 232: A Story of Disaster and Survival*, was turned into a play. People remain fascinated by the horrifying drama and uplifting conclusion.

To find out what qualities these pilots share, I analyzed five commercial flights that went terribly wrong but were saved from total disaster by the actions of the flight crew. I interviewed these pilots about their experiences, asking them what factors had led to the outcome. Their stories had several themes in common and I've grouped them under headings to show that. Innovation was one consistent theme and this should not be surprising. After all, machines do not improvise, and computers are not creative. What pilots bring to the cockpit is their humanity. It is their greatest contribution.

Knowledge and Experience

In the middle of an unseasonably warm winter in Australia, Richard de Crespigny and his adult son Alexander took me boating in Sydney Harbour. Windblown and athletic, the men were clearly in their element; in their nautical shirts and deck shoes, they could have been modeling for the Vineyard Vines catalog.

Then something went wrong with the engine, and de Crespigny had to strip down to his swimsuit and hang on from the fantail to repair the motor so we could get moving. Drenched and oil-spattered, he no longer looked like the man Australia has come to know as Captain Fantastic, but once again his mechanical know-how and experience saved the day.

Call it maturity, knowledge, or time at the wheel, when a pilot gets to a certain age, there's little he or she hasn't seen before. John Gadzinski, a pilot for a U.S. airline and a safety and hazard specialist, says, "You've already been vaccinated as far as your experience and reactions go. It's less and less a deer-in-the-headlights look and more 'Okay, this is what we're gonna do.'"

Still, many experienced airline pilots don't have de Crespigny's in-depth knowledge of the Airbus A380. That's what he was flying on November 4, 2010, when he led a team of five pilots to land a severely crippled airliner with 469 people aboard. For nearly two harrowing hours the jumbo jet circled above the Singapore Strait after an uncontained engine failure blew holes in the wing and fuselage and disabled multiple critical systems.

It was a clear, sunny morning when Qantas Flight 32 departed Singapore's Changi Airport. Then, while passing through seven thousand feet, passengers were jolted by a loud bang. Mike Tooke, seated on the left side of the plane, saw "a flash of white off the inner engine. Then there was an incredibly loud second bang, and the whole plane started to vibrate." Five seconds later, he said, "it felt like we were plunging out of the sky."

From below the wing, a stream of atomized fuel was hosing out of the tank. Some passengers took out their phones and recorded the terrifying sight, no doubt believing they were capturing the last moments of their lives.

On the flight deck, de Crespigny had been about to turn off the seatbelt sign when he heard the two bangs followed by the repetitive beeping of the master warning system. He pushed the altitude hold button, which reduced engine thrust and stress on the engines. It also lowered the nose, which was what caused Tooke to think the plane was "plunging out of the sky."

This simple response wasn't reflex on de Crespigny's part. He was reaching back to an experience he had had nearly a decade before, when he was a passenger on a Qantas 767 that had an engine explode as the plane was ascending after takeoff. In his book *QF32*, he writes that he was impressed by how quickly the 767 captain moderated the plane's violent shaking. Once back on the ground, de Crespigny asked the man what he had done to reduce thrust so quickly and was told, "I just hit the altitude hold button." The small lesson stuck with him. Remembering and deciding to heed it was the first of many decisions, sometimes quick and sometimes after achingly slow deliberation, that contributed to the happy ending of his own near disaster.

My friend David Paqua, a general aviation pilot, once told me, "A pilot can have a thousand hours of experience, or he can fly the same hour one thousand times." Pilots such as de Crespigny use each hour in the air, and even their hours on the ground, to become utterly familiar with the physics and the mechanics of flight. Heck, before QF-32, de Crespigny had visited the factories of both Airbus and Rolls-Royce, gathering material for a technical book he was writing about big jets, including the A380, the plane that gave him such a hard time on that fateful day in 2009.

By contrast, Robert Pearson, the Air Canada captain whom I wrote about earlier in the book, and whose brand-new Boeing 767 would run out of fuel halfway across Canada in July 1983, told me he didn't know enough about the plane he was flying on the day of his near catastrophe, and neither did his airline. "These airplanes came out of Boeing flown by test pilots who had known every rivet and bolt," Pearson said. They arrived at the airline to be flown by "guys like me who knew nothing" about the revolutionary design. His drama began with his not understanding the way the computers assisted the airliner. This complexity masked a very simple problem: the plane was out of fuel. "We didn't know what the problem was. Even when the engines were failing, we were wondering, 'How the hell can computers shut down engines?'" Pearson said.

At the same time, a lifetime of piloting all kinds of planes, including and especially unpowered gliders, enabled Pearson to land the 767 successfully without engines. Each hour in the air can teach something new to even the most experienced pilot.

When the fuel pump alarms started illuminating on Air Canada Flight 143 that summer day in 1983, Pearson said no one on the flight deck had any idea what could be wrong. The engines were still working, and the flight management computers indicated that there was plenty of fuel. Remember, the pilots had manually entered the pounds of fuel loaded, but the 767 flight management system interpreted the input as kilos, a unit of measurement roughly double that of a pound.

As a sign of just how confusing all this was to the crew, Pearson's first announcement to his passengers was that the plane's computer had gone kaput and the flight would divert to Winnipeg to get things sorted out.

When the engines spooled down, the pilots realized they could not spend any more time trying to figure it out. That was in the past. What was going to happen in the immediate future—that was up to them.

Pearson told me that he and the first officer, Maurice Quintal, who died in 2015, needed to focus on how and where they would glide the plane. Pearson was flying, but with only basic instruments. Quintal was doing the math, logging the distance to the closest airfields and comparing it to how quickly the plane was losing altitude.

"I believed we could make it" to Winnipeg, Pearson said, but Quintal's calculations showed otherwise.

While in the military, Quintal had trained at Gimli, an air force base fourteen miles off to the right of where they were flying. They had more than enough altitude to get there—too much altitude, in fact. As the plane approached the airfield, it was too high, and the pilots had no ability to slow it. Quintal lowered the landing gear, but it wasn't enough. So Pearson used a side slip (or crab) maneuver he had honed towing gliders in his off time. Using the rudder, the panels on the wings called ailerons, and the elevators, he turned the fuselage into the airstream so that the plane's bulky metal flank would work against its movement through the air. You can mimic the effect by putting your hand out the window of a moving car with the palm facing forward. You'll feel the resistance right away. That's what Pearson was counting on to help bring down his speed.

"We were using the fuselage as an airbrake," Pearson said. It gave everyone on board a bone-jarringly rocky ride, but it worked.

"I had total tunnel vision. I knew Maurice was beside me. I was one hundred percent concentrated on speed and our relationship to that piece of cement."

In the stories written about the "Gimli Glider," Pearson's

experience in gliders is credited for his inspired innovation that day. Pearson argues the point on two levels. First, the crab maneuver was used most often when he was towing gliders, not flying them. "When coming in on approach on a grass field with a metal rope hanging down [from the plane], you come in high because you don't want to catch the rope on the fence. I'd side-slip every time I came in, and I did a lot of glider towing." Anyway, Pearson points out, gliders have speed brakes, and without power the 767 he was flying did not.

More to the point, he says it was all the flying he did that prepared him for that day. Gliders and airliners, for sure, but also aerobatic planes and ultralights, floatplanes and ski planes on ice and snow—decades of experiences all came flooding back, he told me. "There's something to be gained from everything we do."

Synergy and Teamwork

The philosophy of crew resource management, or CRM, is to merge each pilot's separate strengths to create a more knowledgeable, more experienced team. With de Crespigny on QF-32 were First Officer Matt Hicks and Second Officer Mark Johnson. In what would prove to be fortuitous, two other captains, Dave Evans and Harry Wubben, were also on the flight deck. De Crespigny was being checked out on the A380, and the pilot checking him was being trained as a check captain (that is, learning how to assess whether a pilot meets government criteria). "So," de Crespigny explained, "we had a check captain checking a check captain who was checking me." A total of seventy-six thousand flight hours was represented by the five navy suits.

After reading about the flight of Qantas 32—and I promise, I'll get back to that story shortly—I called Denny Fitch, who told me that all that combined experience would have been an enormous asset for de Crespigny. He should know; he was a hero pilot himself.

In 1989, Fitch was a passenger aboard United Airlines Flight 232 from Denver to Chicago. One hour into the flight, the engine mounted on the tail of the DC-10 came apart at cruise altitude, and a piece of it sliced through a section at the back of the plane, where three separate hydraulic lines came together. Severing the lines caused the fluid to drain, leaving the pilots with no way to turn, slow, or brake the airplane.

Al Haynes was in command of the flight, with First Officer William Records and Second Officer Dudley Dvorak. "Somebody has set a bomb off" was Haynes's first thought when he heard the noise. He was so startled he dropped his coffee.

The pilots were still trying to figure out what happened when they were interrupted by another crisis. The plane began a descending turn to the right. The plane's right wing angled sideways at thirty-eight degrees, far steeper than commercial airline passengers are accustomed to. The DC-10 was on its way to rolling over. Haynes closed the throttle to the left engine and slammed open the lever controlling fuel to the right one. The uneven engine power brought the right wing back up. It was an act of instinct and creativity, gleaned from Haynes's early days flying. He was relying on his knowledge of basic aerodynamics. "You reduce thrust, and that reduces lift," he explained.

Right side up again, the plane began to nose up and down

in a near-constant cycle of ascents and descents called phugoids, which would last throughout the flight. Still, the successful righting of the airplane allowed Haynes to reframe his thinking about what just moments earlier had seemed an impossible situation. The crew could continue to maneuver the airplane using the only control mechanism available: fuel to the engines. Into this scene of spontaneous piloting walked Denny Fitch.

Fitch was a United DC-10 training captain, and he'd gone to the cockpit to see if he could help. He found the men focused on the technique Haynes had just thought up. The added complication was that they could not keep the thrust the same on both engines because that made the plane want to roll over.

"Take one throttle lever in each hand," Haynes told Fitch. "You can do it much smoother than we can." So, positioned between Haynes and Record, Fitch did as instructed. "The throttles became my assignment," Fitch said. None of the four experienced airmen on United 232 had ever tried to fly an airplane this way. No one had ever imagined an airliner losing all flight controls.

Haynes was the commander of the flight, but in the many talks he has given on this event since 1989, he has acknowledged that the skill, talent, and knowledge of all four combined worked to avert complete disaster. "Why would I know more about getting that airplane on the ground under those conditions than the other three?" he said.

When the plane slammed down onto the runway at Iowa's Sioux Gateway Airport three-quarters of an hour later, 185 of the 296 people aboard, including all four of the pilots, survived the crash landing and subsequent fire. One hundred eleven people died, so at best it was a mitigated calamity. It was also a demonstration of the Wright metaphor: a plane otherwise not flyable was wrestled through the air and down onto the runway because control was in the hands of the pilots, and not just any pilots, but a coordinated team whose knowledge, maturity, and experience had a synergetic effect.

Fitch died in 2012, but when I spoke to him about Qantas 32 in the fall of 2010, he reminded me that, as with United 232, those four men on the flight deck that day represented an abundance of hours at the controls of an A380, so it wasn't coincidence that things turned out so well. "You cannot have all the experience in your life to equal seventy-six thousand hours," he said when I told him the combined flight hours of the Qantas crew. The combined flight hours of the pilots on United Flight 232 was even higher: eighty-eight thousand hours. Machines will break, Fitch said, so "at the end of the day it is the human factor that counts."

Decision Making

When the number two engine on de Crespigny's A380 flew apart, the pressure turbine disk fractured into three crescent-shaped pieces, each roughly six feet long and a foot wide. They flew out of the engine like supersize medieval chakrams, taking the back end of the engine cowling with them. Other shrapnel peppered the fuselage and tore holes in the plane's left wing, puncturing the fuel tank and severing a number of wire bundles.

There was no mystery that the problem was with the number two engine, but everything else was uncertain, including why two of the three remaining Rolls-Royce Trent 900 engines were not performing properly. The pilots could not jettison or transfer fuel, and the pumps were not working. Fearful that the last engine might fail, de Crespigny made a request to ATC to climb to ten thousand feet. "I wanted enough altitude so we could glide back to Changi," he reasoned.

Nine months earlier, Captain Sullenberger found himself in a similar situation with even less altitude. He was at three thousand feet following takeoff from New York's LaGuardia Airport when geese flew into the engines, knocking them out. The A320 began a one-thousand-feet-per-minute descent. In his book *Highest Duty*, Sullenberger said he and Skiles knew in less than a minute that they were not going to get to any of the nearby airports. "We were too low, too slow, too far away and pointed in the wrong direction," he wrote. The Hudson River was "long enough, wide enough and on that day, smooth enough to land a jetliner." So he did.

Worrying about whether the Qantas A380 might also turn into a 550-ton glider, de Crespigny calculated just how much altitude he would need to get back to Changi Airport. He wasn't thinking about Sullenberger, he was thinking about the astronaut Neil Armstrong, remembering that when Armstrong was a test pilot flying the X-15 at NASA in the 1960s, he helped develop a technique for gliding the rocket-powered plane back to earth once its fuel was spent.

Armstrong reached altitudes as high as two hundred thousand feet, then glided back to Edwards Air Force Base in California, harnessing gravity to descend in an ever-diminishing spiral. This bit of pioneering aviation was something de Crespigny thought he might need to emulate.

"I was going to do a slow climb to ten thousand feet, to be in gliding zone using the calculation that I could get thirty miles" at that altitude, he explained. His unilateral decision alarmed the other pilots. They wanted to get the plane lower, not higher. For all his charm, de Crespigny is no pussycat. He is opinionated and sometimes stubborn, and he was perturbed that the other airmen did not agree with him. Still, de Crespigny yielded, realizing, as had Al Haynes, that when flying a plane with such grave damage, no one was an expert and everyone was.

"The total number of flight hours accumulated by pilots does not predict the quality of their decisions," Robert Mauro, a professor of psychology at the University of Oregon, wrote in a paper on pilot decision making. "It is experience within a situation that confers expertise." When everyone is a novice, communicating during decision making becomes even more critical.

James Reason describes this as "a willingness on the part of subordinates to speak up and a corresponding willingness on the part of the leader to listen." So concerned was de Crespigny that the three senior pilots on the flight deck not smother the input of the two younger men that he asked the most junior officer, Mark Johnson, to offer his opinions first, followed by his copilot, Matt Hicks.

Using past experience to guide a decision is called associative decision making. And while it can be a fast and effective method, the danger, according to Mauro, is that past experience may not be helpful in "unstable environments or ambiguous situations." Worse still is applying a reflective by-the-book response when creativity or innovation is needed.

As the crew of Qantas 32 flew in circles a mile and a half above the sea, they were consumed by the checklists that were constantly being generated by a computerized airplane trying to diagnose itself and guide the pilots through possible remedies. Fuel was draining overboard from the hole in the left wing tank, and this created several fuel imbalances. When the checklist for wing imbalance appeared, it called for the pilots to open the valves to send fuel from the good tank to the one that had been breached.

"Should we be transferring fuel out of the good right wing into the leaking left wing?" de Crespigny asked his crew. "No," they replied. Many airlines expect pilots to follow standard procedures strictly. Determining when to follow and when to ignore them requires knowledge, experience, logic, mindfulness, communication, and strength, but decision-making strategies are still evolving.

In *The Pilot's Handbook of Aeronautical Knowledge*, the FAA uses the mnemonic 3P, for *perceive*, *process*, and *perform*, to help pilots remember what steps ought to precede a decision. At Lufthansa, the cadets learn a different acronym, FORDEC, for *facts*, *options*, *risks*, *decisions*, *execute*, and *check*. That final C could also stand for *circle back* because the big lesson for pilots is that a decision isn't made and done; it's an ongoing cycle.

Captain Norhisham opted to keep the autothrottles engaged while maneuvering MH-124 back to Perth, but he had to revise that plan because of the constant revving and powering back of the engines. That's one example of reviewing a decision. On Air Canada Flight 143, Pearson and Quintal wanted to land at Winnipeg Airport because emergency equipment would be available and big-city hospitals were nearby. But it was too far away. They considered ditching the plane in Lake Winnipeg, but as Quintal continued updating his distance calculations, he realized they could glide to Gimli. The constant revising of the plan continued, leading to the innovative piloting that has made

Air Canada Flight 143 one of aviation's most talked about recoveries.

Airmanship

The day James Reason gave his presentation on heroic recoveries to the attendees of the conference on risk in 2009, he shared the stage with forty-five-year-old British Airways captain Peter Burkill. Burkill's crash landing in London the year before was the darkest swan among the flock because the 3Ps, FORDEC, CRM—all those intended-to-be-helpful alphabet formulas—were irrelevant. Burkill and First Officer John Coward were faced with a failure so sudden they had only seconds to react.

The pilots and the relief first officer, Conor Magenis, were at the tail end of the ten-hour flight of British Airways 38 from Beijing. As the plane flew over the outskirts of London, Burkill had no worry bigger than whether the gate would be available when they arrived at the airport. The captain could see where they would touch down, on runway 27L, off to the west, on the other side of the borough of Hounslow. Less than a minute before landing, Coward, who was flying the leg, said suddenly, "I can't get any power."

"I remember looking at his hands on the throttles, and I could see the demand: the autothrottle was fully forward," Burkill told me. He was still processing what was going on when Magenis chimed in from behind them, saying it looked like a double engine failure.

"I remember every second of that event. It seemed like the event was three minutes long," Burkill said. In truth, the time that passed from Coward's stunning discovery to the plane hitting the ground was just thirty seconds.

After realizing they would have to make a landing without power, Burkill first decided to leave Coward flying the plane while he concentrated on their options. Ahead loomed frightening obstacles: factories, multistory residences, the Hatton Cross tube station, and a gas station, all of which they would have to fly over to reach the airport. The airfield was surrounded by a high perimeter fence, and on the other side of that, a nine-foot-tall lineup of antennas and airport lights would block a too-low approach.

Burkill eyed the gauges showing ten tons of fuel still in the tanks and worried about fire. Again he checked the throttles, but there was no improvement. Then the first warning horn sounded in the cockpit. The autopilot had kept the flight on the approach path, but the lack of fuel to the engines was causing the plane to slow. The yoke started to vibrate in Coward's hands, and an audible airspeed warning sounded. The rate of descent increased until it was more than double the normal seven hundred feet per minute.

"I knew what it was supposed to feel like, and it was not this," Burkill said. The only way to keep the plane from slamming into the football-size two-story building dead ahead was to reduce the airplane's drag. The landing gear, which had been lowered before the crisis began, was slowing them down, but there wasn't enough time to retract it. Burkill thought it might also help absorb the impact of the inevitable crash landing. He kept thinking.

Before the trouble began, in preparation for the approach, Coward had asked Burkill to set the flaps to thirty degrees.

This makes the wing more comma shaped so the plane can fly at slower speeds. Burkill now considered whether slightly flattening the curved surfaces might be enough to keep the plane flying. He reached over to the flaps lever and, after a moment's hesitation, moved it from thirty to twenty-five degrees. He did not consult Coward; there just wasn't time. The effect was immediate: the descent slowed.

Gus Macmillan, a musician from Melbourne, was in a window seat just behind the wing on the right. "I remember thinking, 'We only just cleared that fence' as the grass of the runway unfolded beneath us," he told the Australian newspaper *The Age*. The plane hit the grass field 890 feet short of the runway, at 124 miles per hour.

Burkill's decision had given the plane an additional 164 feet in the air and enabled Flight 38 to fly over the ominous white building, pass the gas station, clear the highway, and cross safely over the imposing electrified metal barrier of antennas and runway lights. All 152 people on board survived.

"I wish I had time to actually communicate" with the others, Burkill told me later. A believer in the benefits of CRM, he says that not only did they face a situation for which they were never trained, but also there was no time to use any of their crisis management tools. It was nothing like the sessions in the simulator with all that time to talk to ATC and the cabin crew, glorious minutes to consider options with others on the flight deck. In a real-life emergency, he had to make everything up as he went along.

"I'm in this gray area," he said, "this gray area that no pilot wants to be in, with no checklists for my situation and nothing written down." In the modern jet age, the loss of all engines is so exceedingly rare that it is not a scenario practiced by pilots in their simulator sessions.

Without diminishing the horrifying experience, one that leads many hero pilots to wrestle with posttraumatic stress disorder long after the public accolades subside, this lack of guidance is where humans excel.

Uncertainty and Surprise

If you are wondering if I've forgotten you over the Singapore Strait in a noisy and unstable jumbo jet, flying a horse track holding pattern with the 469 frightened travelers on Qantas Flight 32, I have not. I left you to experience just a portion of the hour and forty-five minutes during which the pilots and their passengers flew on, uncertain of their fate.

One of the mighty plane's four engines was out, and two others were degraded. Imagine Second Officer Mark Johnson walking down the aisle of the airplane, straining to assess the damage by looking out the passengers' windows.

Perhaps, like me, you wonder why the pilots didn't just put that airplane back on the ground ASAP? This was a subject of discussion in the cockpit.

De Crespigny said, "We reconsidered this option every fifteen minutes in the air. This was not the time to panic and make irrational decisions. We were on a fact-finding journey; we had to understand how much of the A380 we had left before we could hope to land."

Yes, flying was a hazard. Still, before committing to landing a plane that was still too heavy because of all the fuel loaded for the flight to Sydney and only 65 percent of its braking power, de Crespigny had some questions he wanted answered. Could any of their many problems be fixed? How would those they could not solve affect their ability to land? They could not know the answers. Too much was wrong with the airplane.

Nearly an hour into the flight, Dave Evans and Harry Wubben set about calculating how much pavement the plane would need to stop. The longest runway at Changi Airport was 13,000 feet. Evans and Wubben calculated that the plane would need 12,700 feet of it, more than twice that for a normal landing. It was achievable, albeit with a very tight margin. That was good news.

When the wheels of the giant A380 touched the ground, de Crespigny jammed both feet on the brakes, and miraculously the plane slowed, stopping within the distance calculated by Evans and Wubben. Yet the drama was still a long way from over.

Fuel continued to dump out of the wing, but now it was pooling within range of the brakes that had just done all that stopping and were heated to sixteen hundred degrees Fahrenheit. The fire trucks could not approach the aircraft until the engines were powered down, but when the crew switched them off, the plane was thrown into darkness. Nine of the ten cockpit display screens had failed. Six of the seven radios were dead, and the number one engine kept on spinning.

For nearly an hour, everyone sat in the dark and sweltering airplane, while the crew worried about the ignition of the fuel and the threat posed by an engine that still whined as if it were in the air. Dave Evans told the Royal Aeronautical Society, "We've got a situation where there is fuel, hot brakes, and an engine that we can't shut down. And really the safest place was on board the aircraft."

It took three hours of dousing the engine with water and foam to finally get it to stop turning.

As with the other dramatic events described in this chapter, the pilots knew the uncertainty would come to an end—in hours for Qantas Flight 32, forty-five minutes for United Flight 232, half an hour for Malaysia Flight 124, a quarter of an hour for Air Canada Flight 143, and less than a minute for British Airways Flight 38. They did not know what the end would look like. It's easy to forget that.

All the pilots I've written about here were deliberate in their decisions, and they struggled to improve the odds weighing heavily against them—and they all had to contend with an eleventh-hour disruption. It was an engine that would not shut down and the ongoing risk of fire on Qantas 32; and it was a wind shear alarm that rattled Norhisham and Foong as the Malaysia Airlines 777 approached Perth. With United Flight 232, as the DC-10 approached Sioux Gateway Airport, another of the phugoids sent the jet plummeting just three hundred feet above the runway.

"That's where our luck ran out. We just ran out of altitude, trying to correct it," Haynes told a NASA conference on risk in 1991. "That close to the ground, we didn't have time." The right wing and tail section broke off, a fire erupted, and the body of the airplane bounced along the runway and broke apart. Most of the fatalities were in the back of the plane and in the first-class cabin behind the cockpit, which broke away during impact.

The most bizarre postscript came in the final moments of the Air Canada flight that came to be known as the Gimli Glider.

The nose gear collapsed as the plane touched down, and the front end of the plane hit the ground with "a hell of a thump," as Pearson recalled it. The dragging of the metal across the pavement acted to brake the plane, which was a blessing. For, unseen by either of the pilots as they approached what they thought to be an abandoned airfield, the former air base had been converted into a motorsports track. On that day, it was being used by the Winnipeg Sports Car Club and was crowded with spectators enjoying the summer afternoon. Miraculously, no one on the ground was injured.

After wrestling with perplexing calamities in the air, not getting a break in the final seconds just seems wrong, but hurdles right up to the end do not surprise retired airline pilot and air safety specialist John Cox. "Air accidents are complex," Cox explained. "In some cases, factors outside the original problem can come into play, and in others, there is so much going on that it is not possible to anticipate all the possibilities."

The Opposite of Despair

Norhisham Kassim praised God after his harrowing brush with disaster, and Al Haynes chimed in with his belief that "something guides us in all we do," adding that strength can be found by looking inward, a philosophy shared by Pearson and Burkill. This last component of resiliency is what James Reason calls "realistic optimism," the opposite of despair, a stubborn belief that things will be all right in the end.

"You have to believe in yourself. Every time you go to work you're doing something that not everyone on the street can do," Burkill told me. In managing emergencies, confidence is necessary, "for sure," he said. Pearson chimed in with the circular argument that experience makes a pilot confident, and confidence can lead to positive outcomes.

"Pilots should feel they can handle anything," Haynes told me. "If you don't have that feeling, you shouldn't be flying."

These nerve-racking flights are rare. The general public will never know just how often pilots avert disaster much earlier in the chain, but several airline executives say that safety threats are interrupted all the time.

Despite the ambiguities in how the increasingly complex airliner affects the pilots' ability to interact with it, one thing is clear: the amount of data available from new-generation airplanes is a remarkable tool. Hundreds of details on every routine flight are collected and analyzed as airlines try to determine how often their operations veer outside the safety envelope. Voice and data recorders offer an after-the-fact view, but information collected during normal trips can be downloaded, combined with others, and analyzed in order to discover hidden weaknesses in maintenance, training, or operational procedures.

"Even with a good outcome, every part of the story isn't perfect," said Billy Nolen, a former captain with American Airlines and now senior vice president of safety for the U.S. airline trade association Airlines for America. Reviewing large numbers of flights allows a carrier to understand how close to the edge uneventful flights get. "What is our data showing? What is our story?"

It's not exactly studying the silver linings instead of the clouds. Much more can be done to get there, according to Captain Fantastic (a.k.a. de Crespigny), who has taken on the study of human achievement with the same energy he has devoted to flying. "Many things improve when we mine the big data for successes," he said. From the few hero pilots who have accomplished dramatic saves to the many who overcome hurdles and safely bring their passengers to their destinations—these examples should be examined for the lessons they hold. "We'll be able to change our definition of safety from avoiding what goes wrong to ensuring things go right," de Crespigny said.

On a summer morning in 2006, Capt. Cort Tangeman and First Officer Laura Strand were approaching Chicago's O'Hare Airport. They'd taken an MD-80 airliner on the overnight flight from Los Angeles. Strand was flying the leg, and she called for Tangeman to lower the landing gear as they neared the airport, but Tangeman found that the nose gear doors would not open.

"At that point we had been up all night, and it's kind of like shock and awe," Tangeman said. "You're not sure what you're seeing, and you know it is not going to be solved between now and the time you land." The crew took the plane down to about five hundred feet, making an unusual flyby right down the runway of one of the world's busiest airports. From there, a tower controller eyeballed the plane using binoculars. This confirmed that while the main gear was down, the wheels under the plane's nose were not.

The crew was given an area off of the approach path to fly while they planned for landing. They were not immediately concerned about fuel, but soon they would be.

"I was alarmed at how much fuel we were burning, because flaps were down in early-approach mode setting, and the gear was down and we were low. Airplanes burn a lot of fuel at low altitude," Tangeman said. He had taken over the flying duties from Strand, who was now working the radios. "When that fuel light went on, that added another level to that event."

The pilots had talked to maintenance and tried to extend the gear manually, without success. Tangeman remembers the tension and finality that came with the realization that the emergency was real and that all their skills would be required to avert disaster.

"This is not a simulator, we can't step back and do it again; we're fuel low," he said he thought at the time. "We're not getting out of this."

As the plane touched down, Tangeman decided to stay on the main gear as long as possible without operating the thrust reversers. "When the aluminum skin of the MD-80 finally hit the tarmac, the sound was like running a Skil saw on a garbage can, and we stopped really fast at the seventy-five-hundred-foot mark, fully loaded, with no reversers." It was a landing so flawless that none of the 136 people aboard was injured, and damage to the plane was minimal.

A television news helicopter equipped with a camera had recorded the last thirty seconds of the landing, providing a riveting element to an already compelling story. Maybe this is the reason Tangeman's and Strand's performances that day became the subject of human factors training at American for the next eighteen months.

Tangeman was encouraged to share his story with his fellow pilots, and the subject to which he kept returning was the value of the lessons he'd learned from others. "There are no top guns" in the airline cockpit, he said. Senior pilots who share their expertise enable an atmosphere where pilots routinely save the day.

"The most influential thing in my life has been working with other great captains," Tangeman told me. "Nothing replaces great mentoring."

At a time, and in an industry, in which automation is preferable to the human touch, humanity's marvelous flip side is often overlooked and underappreciated, except in cases of hero pilots such as those you've read about here. Their stories are uplifting, but they are certainly not the only people contributing to the complex system that keeps flying safe. Aircraft and engine designers, airline workers and maintenance engineers, air traffic controllers and regulators—they all play a role; as do passengers when they note the closest emergency exit to their seat and keep their belts fastened throughout the flight.

When things go wrong, as they inevitably do, the crash detectives find the lessons in catastrophe. Our uniquely human ability to learn from mistakes, to think, create, and innovate, works better than we will ever know.

ACKNOWLEDGMENTS

Writing a book is an exercise in patience—not for me, but for everyone whose path has crossed mine over the past two years. I pestered them all. People I knew and people I didn't. An astonishing number of them provided assistance.

Some are quoted or profiled in this book. Others gave background help, research, guidance, fact checking, and provocation, all of which crystallized my thoughts. Without all these people, this book would be incomplete.

I am deeply grateful for the knowledgeable people mentioned here and also for those who, fearing negative consequences, asked that I not thank them by name. Their contributions infuse every page.

In the digital age, a librarian's job is constantly evolving. I had the opportunity to work with two stellar examples of this still-vital profession, Yvette Yurubi, who presides over nearly seven decades of Pan American Airways history at the University of Miami, and Nick Nagurney, from the Perrot Memorial Library in my hometown of Old Greenwich, Connecticut, who cheerfully searched the world for some very obscure titles. To the authors of the books and reports in my bibliography, thanks for making complicated subjects comprehensible.

Sometimes the written word isn't enough, so I relied on Bob Benzon, James Blaszczak, Mike Bowers, Barbara Burian, John Cox, Key Dismukes, Olivier Ferrante, Peter Fiegehen, Pete Frey, John Gadzinski, Darren Gaines, Mitch Garber, Keith Hagy, Tom Haueter, Guy Hirst, Kevin Humphreys, Judy Jeevarajan, Jim Karsh, Rory Kay, Lewis Larsen, John Lauber, Robert MacIntosh, John Nance, Michael O'Rourke, Kazunori Ozawa, David Paqua, Mike Poole, Helena Reidemar, Eduard M. Ricaurte, Donald Sadoway, Steve Saint Amour, Gary Santos, Ron Schleede, Patrick Smith, and Robert Swaim.

Working in an unfamiliar culture is always challenging. Providing assistance in Malaysia in too many ways to count were Maureen Jeyasooriar, Riza Johari, and Anita Woo. Although, like all Malaysians, they were dazed by the disappearance of Malaysia 370, these women worked with energy and dedication.

In Japan, Takeo Aizawa started off as my legman, became my translator, moved into the position of researcher, and then adviser, and will always be a dear friend.

For digging up old records and sharing stories of events from long ago, special thanks to Ed Dover of Albuquerque, and the late Nick Tramontano. Others deserving of a special shout-out are Stuart Macfarlane and Anne Cassin in New Zealand, Mick Quinn and Ben Sandilands in Australia, Samir Kohli in India, George Jehn in New York, Les Filotas in Ottawa, Guy Noffsinger in Washington, DC, Jeff Kriendler in Miami, and Graham Simons and Susan Williams in England.

Officials with the following organizations went out of their way to accommodate me in one way or another: Daniel Baker,

CEO at FlightAware; Perry Flint of the International Air Transport Association; and Markus Ruediger of Star Alliance. From Lufthansa, Matthias Kippenberg, Nils Haupt, and Martin Riecken (the latter two employed elsewhere now). Also thanks to Corey Caldwell of the Air Line Pilots Association; Martin Dolan, now retired from the Australian Transport Safety Bureau; Robert Garner of the High Altitude Chamber Lab at Arizona State University, Mesa; Mary Anne Greczyn at Airbus; Peter Knudson of the National Transportation Safety Board; James Stabile and James Stabile Jr. of Aeronautical Data Systems; and Mamoru Takahashi of the Japan Transport Safety Board.

Special thanks to ABC News for bringing me in to help with the network's coverage when air disasters happen. While I was in Malaysia, it was a gift to work with Mike Gudgell, Matt Hosford, David Kerley, David Reiter, Gloria Rivera, Brian Ross, Rhonda Schwartz, Ben Sherwood, Jon Williams, and Bob Woodruff.

Exercising patience is one thing, but there's little time for real exercise when writing a book. Thanks to Joanna Stark at Rebel Desk for enabling me to write on my feet; Steven Fiorenza of Advanced Physical Therapy of Stamford for stretching out my kinks; and M. J. Kim of Kida NYC for helping me feel pretty.

The two best critics in the world love me, if not everything I write. My sister Andrea Lee Negroni is a lawyer, but she slices sentences with such precision she could have been a surgeon. My husband, *New York Times* editor Jim Schembari, is an accomplished word polisher, but he had his work cut out for him in tackling mine.

I hope their efforts made my copy a little easier to handle for my editors, Shannon Kelly and Meg Leder, and the very clever Emily Murdock Baker, formerly of Penguin and now the head of EMB Editorial. Special thanks to them and to publicist Christopher Smith and to my agent, Anna Sproul-Latimer of Ross Yoon, who is prone to saying, "What can I do to help?" just when I need it most. Anna and I might never have met if not for the situational awareness of the beautiful and talented Dara Kaye, also of Ross Yoon. Thanks to my research assistant Chrissi Culver, whose tracking-down and following-up skills will surely be applied to the benefit of the flying public in her new position as an air traffic controller.

Women are a small part of the aviation geek community, but our numbers are growing. I'm so thankful that Chrissi, Emily, and Anna are among them.

Unending appreciation and love to my family: my husband, Jim, and my children, Antonio, Sam, Joseph, Marian, and her husband, Elliot Speed, for all their support. My undying gratitude to God for saving grace.

BIBLIOGRAPHY

Adair, W. *The Mystery of Flight 427: Inside a Crash Investigation.* Washington, DC: Smithsonian Institution Press, 2002.

Air Accident Report No. 79-139. Air New Zealand McDonnell Douglas DC 10-30 ZK-NZP. Ross Island Antarctica 28 November 1979. Wellington: Office of Air Accidents Investigation, Ministry of Transport, 1979.

Bainerman, J. *The Crimes of a President: New Revelations on Conspiracy & Cover-up in the Bush & Reagan Administrations.* New York: Shapolsky, 1992.

Bartelski, J. *Disasters in the Air: Mysterious Air Disasters Explained.* Airlife, 2001.

Beaty, D. *Strange Encounters: Mysteries of the Air.* New York: Atheneum, 1984.

Booth, T. *Admiralty Salvage in Peace and War 1906–2006: Grope, Grub and Tremble.* Barnsley, England: Pen & Sword, 2007.

Bragg, R. L. "Tenerife—A Survivor's Tale". *Flight Safety Australia*, September-October 2007.

Brock, H. *Flying the Oceans: A Pilot's Story of Pan Am, 1935–1955.* Lanham, Maryland: Jack Aronson, 1978.

Burkill, P., and M. Burkill. *Thirty Seconds To Impact*. Bloomington, Indiana: AuthorHouse, 2010.

Butler, S. *East to the Dawn: The Life of Amelia Earhart*. Boston: Addison-Wesley, 1997.

Choisser, J. P. *Malaysia Flight MH370—Lost in the Dark: In Defense of the Pilots: An Engineer's Perspective*. CreateSpace, 2014.

Crouch, G. *China's Wings*. New York: Bantam, 2012.

Davis, J. R., et al., eds. *Fundamentals of Aerospace Medicine*, 4th edition. Philadelphia: Lippincott Williams & Wilkins, 2008.

de Crespigny, R. *QF32*. Sydney: Pan Macmillian Australia, 2012.

deHaven-Smith , L. *Conspiracy Theory in America*. Austin: University of Texas Press, 2014.

Filotas, L. *Improbable Cause: Deceit and Dissent in the Investigation of America's Worst Military Air Disaster*. BookSurge, 2007.

Gawande, A. *The Checklist Manifesto: How to Get Things Right*. New York: Metropolitian Books, 2009.

Gero, D. *Aviation Disasters: The World's Major Civil Airliner Crashes Since 1950, 4th ed*. Stroud, England: Patrick Stephens, 2006.

Gonzales, L. *Flight 232: A Story of Disaster and Survival*. New York: W. W. Norton, 2014.

Griffioen, H. *Air Crash Investigators: The Crash of Helios Airways Flight 522*. Lulu.com , 2009.

Haine, E. A. *Disaster in the Air*. Cranbury, New Jersey: Cornwall Books, 2000.

Hill, C. N. *Fix on the Rising Sun The Clipper Hi-jacking of 1938 – and the Ultimate M.I.A.'s*. Bloomington, Indiana: 1st Books Library, 2000.

Hoffer, W., and M. M. Hoffer. *Free Fall: A True Story*. New York: St. Martin's Press, 1989.

Holmes, P. *Daughters of Erebus*. Auckland: Hodder Moa, 2011.

Inkster, I., ed. *History of Technology 2005*, volume 26. London: Continuum, 2006.

Jackson, R. *China Clipper.* Everest House, 1980.

Jehn, G. *Final Destination: Disaster: What Really Happened to Eastern Airlines.* Howard Beach, New York: Changing Lives Press, 2014.

Keith, R.A. *Bush Pilot With a Briefcase: The Incredible Story of Aviation Pioneer Grant McConachie.* Vancouver, Canada: Douglas & McIntyre, 1972.

Kemp, K. *Flight of the Titans: Boeing, Airbus and the Battle for the Future of Air Travel.* London: Virgin Books, 2006.

Kohli, S. *Into Oblivion: Understanding #MH370.* CreateSpace, 2014.

Langewiesche, W. *Fly by Wire: The Geese, The Glide, The Miracle on the Hudson.* London: Picador, 2010.

Levine, S. *The Powerhouse: Inside the Invention of a Battery to Save the World.* New York: Viking, 2015.

Lindbergh, C. A. *Of Flight and Life.* New York: Charles Scribner's Sons, 1948.

Long, E. M. and M. K. Long. *Amelia Earhart: The Mystery Solved.* New York: Simon & Schuster, 1999.

Mahon, P. *Report of the Royal Commission Crash on Mt. Erebus.* Wellington, New Zealand: Hasselberg Government Printer, 1981.

Mahon, P. *Verdict on Erebus.* London: Collins, 1984.

McCain, J., ed. *Aviation Accident Investigations: Hearing Before the Committee on Commerce, Science, & Transportation, U. S. Senate.* Darby, Pennsylvania: Diane Publishing, 1997.

McCullough, D. *The Wright Brothers.* New York: Simon & Schuster, 2015.

Medina, J. *Brain Rules.* Seattle: Pear Press, 2014.

Micklos, J. *Unsolved: What Really Happened to Amelia Earhart.* New York: Enslow, 2006.

Murphy, J. D. *Courage to Execute: What Elite U.S. Military Units Can Teach Business About Leadership and Team Performance.* New York: Wiley, 2014.

Reason, J. *The Human Contribution: Unsafe Acts, Accidents and Heroic Recoveries*. New York: Ashgate, 2008.

Reed, T. and D. Reed. *American Airlines, US Airways and the Creation of the World's Largest Airline*. Jefferson, North Carolina: McFarland, 2014.

Serling, R. J. *The Jet Age*. New York: Time Life Books, 1982.

Simons, G. *Comet!: The World's First Jet Airliner*. Barnsley, England: Pen & Sword Aviation, 2013.

Soucie, D. *Malaysia Airlines Flight 370: Why it Disappeared—and Why It's Only a Matter of Time Before This Happens Again*. New York: Skyhorse, 2015.

Sullenberger, C. with J. Zaslow. *Highest Duty: My Search for What Really Matters*. New York: William Morrow, 2009.

U.S. Department of Transportation, Federal Aviation Administration. *The Pilot's Handbook of Aeronautical Knowledge 2008*. Washington, DC: GPO, 2008.

Vette, G., and J. MacDonald. *Impact Erebus*. Lanham, Maryland: Sheridan House, 1983.

Wagner, A.H. and L.E. Braxton. *Birth of a Legend: The Bomber Mafia and the Y1B-17*. Bloomington, Indiana: Trafford, 2012.

Walker, J. *The United States of Paranoia: A Conspiracy Theory*. New York: Harper Perennial, 2014.

Wecht, C. H. and M. Curriden. *Tales from the Morgue: Forensic Answers to Nine Famous Cases*. Amherst, New York: Prometheus Books, 2005.

Wiggins, M. W. and T. Loveday, eds. *Diagnostic Expertise in Organizational Environments*. New York: Ashgate, 2015.

Williams, S. *Who Killed Hammarskjöld? The UN, the Cold War and White Supremacy in Africa*. New York: Oxford, 2014.

Wise, J. *Fatal Descent*. Amazon Digital Services, 2015.

INDEX

INDEX

INDEX

INDEX

INDEX